Since 1956, *New Scientist* has established a world-beating reputation for exploring and uncovering the latest developments and discoveries in science and technology, placing them in context and exploring what they mean for the future. Each week through a variety of different channels, including print, online, social media and more, *New Scientist* reaches over 5 million highly engaged readers around the world.

D1369155

THE BRAIN

Everything You Need to Know

ALISON GEORGE

NewScientist

First published in the United States of America in 2018
by Nicholas Brealey Publishing
First published in Great Britain in 2018 by John Murray (Publishers)
An Hachette UK company

This paperback edition published in 2022

7

Illustrations by Valentina D'Efilippo 2018
Additional writing by Helen Thomson,
Caroline Williams and Graham Lawton
Fact checking by Chris Simms

A CIP catalogue record for this title
is available from the British Library

Paperback ISBN 978-1-529-36331-9
UK eBook ISBN 978-1-473-62931-8
US eBook ISBN 978-1-473-68507-9

Typeset in Celeste by
Palimpsest Book Production Ltd, Falkirk, Stirlingshire

Printed and bound in Great Britain by Clays Ltd, Elcograf S.p.A.

John Murray policy is to use papers that are natural, renewable and
recyclable products and made from wood grown in sustainable forests.
The logging and manufacturing processes are expected to conform
to the environmental regulations of the country of origin.

John Murray (Publishers)
Carmelite House
50 Victoria Embankment
London EC4Y 0DZ

Nicholas Brealey Publishing
Hachette Book Group
Market Place Center, 53 State Street
Boston, MA 02109 USA

www.johnmurraypress.co.uk
www.nbuspublishing.com
www.newscientist.com

CONTENTS

13. TROUBLESHOOTING

14. UNLOCKING YOUR POTENTIAL

INTRODUCTION

IF YOU'RE READING this, congratulations! You're the proud owner of the most complex information-processing device in the known universe. An adult human brain weighs in at just 1.4 kilograms but packs an incredible punch: 86 billion nerve cells with nearly a trillion connections between them. This biological supercomputer comes equipped with all sorts of design features – from consciousness and memory to intelligence and creativity – but also has many bugs and weaknesses. The problem is, you don't get a user's manual. You have to just plug and play. As a result, most of us never properly understand how our brains work and what they're truly capable of.

Perhaps that's no surprise, because your brain is like an iceberg. Think of the visible portion above the waterline as your conscious awareness, and the rest as your unconscious. This submerged 90 per cent is where most of the action takes place, but it is extremely adept at hiding what it's up to.

Take your perceptions of the world around you. It's easy to assume that your eyes act like a video camera, faithfully recording what is going on in the outside world and relaying it in glorious 3D technicolour to your brain. In reality, most of what you see is made up by your brain, a carefully orchestrated hallucination of reality.

This is easy to demonstrate. Close your left eye and look hard to the left. You will notice a large, fleshy, blurry object; this is your nose. It is always within your field of vision, so why don't you see it all the time? It is because your brain has decided it isn't important information and edits it out. In fact, your eyes are constantly darting around all over the place, but you're not aware of this either, because your brain operates sophisticated software to link these fragmentary snapshots into a seamless movie. What you 'see' is a largely a fabrication by your mind.

There are dozens of similar easy-to-do experiments throughout this book, often demonstrating how what we are consciously aware of diverts radically from reality.

Take your earliest memory. Mine is visiting my new sister in hospital soon after my second birthday. I can vividly picture her lying asleep in a cot wrapped in a blanket, me on my mother's bed and a woman knitting on the other side of the hospital room. But I'm aware that this is unlikely to be a true recollection. Although a few people can recall events from their third year, the average age of a first memory is around three-and-a-half – and some people can't recall anything before they were six.

It's possible I could be a childhood memory prodigy, but I doubt it. Through my work at *New Scientist* I've had the privilege of interviewing false memory expert Elizabeth Loftus, who has demonstrated how easy it is to create 'memories' of things that didn't happen. Though it feels very real to me, I now think that my earliest memory is likely to be something my brain invented at a later date, based on other people's anecdotes and photographs.

This raises many questions: why do we remember virtually nothing from early childhood, when so many exciting things are going on? Why do only certain events stick in our minds? Fascinating research into how memory works shows that its purpose might be to imagine the future rather than recall the past (see What if we couldn't forget?, page 188, for much more on that).

This kind of counterintuitive discovery is why we decided to put together a manual to help you better understand the workings of your own mind. As neuroscientists develop the technology to delve ever more deeply into our grey matter, many of the certainties we feel begin to break down. It is even possible that our sense of being conscious and of having free will could be illusions (see What if you have no free will?, page 239, for more on this).

The Brain: Everything You Need to Know doesn't just shine light on what's going on inside your head – it also shows you ways to improve it too. You will discover how to use brainpower to defy ageing, improve your memory and upgrade your dreams. There's even a scientifically proven guide to futureproofing your brain.

This book also contains many experiments you can try on yourself: quizzes to test your memory skills, how to hallucinate without the aid of mind-altering drugs, and even discover whether you are a psychopath.

Unlocking how the brain is built and how it operates is a crucial part of what we do at *New Scientist*. This book is a testament to the enquiring minds and brainpower of my incredible colleagues and the magazine's writers, covering as it does a vast range of topics, from the nature of thought to the brain farts that make you feel stupid.

'My brain is open' was the greeting used by the renowned mathematician and eccentric Paul Erdös. It was an approach that served him well: he is regarded as one of the greatest mathematical minds of the twentieth century. You too will have to do something similar to get your head round some of the mind-bending concepts in this book.

Go on, open your brain.

Alison George
April 2018

01.

MEET YOUR BRAIN

WELCOME TO YOUR BRAIN

THE BRAIN IS an incredible organ. So incredible that we barely notice its superpowers in action.

Picture this: you have a pile of washing-up in front of you. It wobbles precariously as you balance another saucepan on its summit. For a second it looks like the whole stack will come down. But it doesn't. Swiftly, instinctively, you save it.

Congratulations – not just on another domestic disaster averted, but also on showing a peculiarly human genius. No other species can perform complex, real-time calculations of their physical environment quite like the ones that rescued your washing-up. And you didn't even notice your brain doing the hard work.

This kind of advanced physics is just one of your brain's effortless talents. It is also a time machine and a crystal ball, zooming back decades to relive a memory stored long ago or predicting scenarios far into the future.

All this is done at lightning speed. In less than the blink of an eye it can scan multiple pictures to identify the right one. Yet its power consumption would make a computer engineer weep with envy. The human brain runs at around 20 watts, less than a typical light bulb. Contrast this with AlphaGo,

Google's computer program which in 2016 beat Korean grand-master Lee Sedol at the fiendishly complicated game Go. While AlphaGo may have been victorious, it required around 50,000 times more energy than Sedol's brain to work out the moves, and a sophisticated cooling system in order not to overheat. Sedol's brainpower barely caused him to break into a sweat.

All this from a wrinkled, pinkish-grey 1.4-kilogram blob with the consistency of tofu.

The brain is the most confusing, complicated and arguably the ugliest organ in our body. It produces every thought, action, memory, feeling and experience of the world. Over the course of history it has enabled our species to build civilisations, create art and fly to the moon.

Brains carry out a remarkable range of feats, but their fundamental purpose is really quite simple: to allow their owners to react to changes in the environment. They enable us to move away from danger, remember a distant food source, to distinguish friend from foe.

Contrast this with a brainless plant, which must stay put and endure the conditions that nature throws at it. Indeed, some scientists argue that the real reason for having a brain is not to perceive the world or to think, but to move. The clinching evidence comes from the larvae of sea squirts. These animals have a simple brain, but once they permanently settle down on a rock it degenerates and is absorbed into the body. Brainpower is no longer needed.

Brain vs heart

In times past, the wonders of brains were not obvious to their owners. The ancient Egyptians famously removed the brains of bodies preserved as mummies by inserting a hook through the nose, as the organ wasn't regarded as important in the afterlife. Hearts were often left in place, as they were believed to be the seat of the person's intelligence and feelings. It took thousands of years to develop a full understanding of our grey matter, and it was only at the dawn of the twentieth century that the building block of the brain – the neuron – was discovered.

We now know that the brain's immense power is derived from the way these cells are connected. A complex ballet of electrical and chemical activity in the brain's network of 86 billion neurons allows us to sense our surroundings, feel, taste and remember. This network is made up of a staggering 170,000 kilometres of nerve fibre. That's almost half the distance from Earth to the moon.

Complex networks

The complexity of the connections between these cells is mind-boggling. Each neuron can make contact with thousands or even tens of thousands of others. Our brains form a million new connections every second of our lives. The pattern and strength of the connections is constantly changing and it is in this network that memories are stored, habits learned and personalities shaped, by reinforcing certain patterns of brain activity and losing others. In fact, the memories you form while reading this article mean that you will have a different brain from the one you started reading it with.

Modern neuroscience has built a sophisticated picture of how our brain works, but there are still many unsolved problems. What is the neural basis of personality? How does the brain process time? And the biggest mystery of all: where does consciousness arise from?

It's possible that we will never crack these hard problems. Perhaps the mind is beyond human understanding? The Nobel prize-winning physicist Erwin Schrödinger certainly thought so when he was pondering the issue in his 1944 book *What is Life?*. The debate continues to this day. In the meantime, more of the brain's genius is being uncovered, from the profound to the mundane: the awesome power of the unconscious mind, the weird sensory mix-up of synaesthesia, the origins of 'aha' moments when everything suddenly clicks – and the reason why it's so hard to stop biting your nails.

It's amazing to think of all the things an ugly lump of grey matter can do.

CAN YOU REALLY BE JUST A BRAIN?

All this talk of neurons and wires can seem a little unnerving. Can our hopes, loves and very existence really be just the outcome of electricity moving through a mass of grey tissue? Back in the seventeenth century, the philosopher René Descartes set the tone in this debate by proclaiming that the brain and mind are cut from very different kinds of cloth. The brain is made of matter, he said, but the mind, with our thoughts, beliefs, mental lives and memories is immaterial – something that cannot be seen, touched or observed. Today's neuroscientists disagree with Descartes' mind–brain dualism. To them, the mind isn't a special non-physical substance, but just the manifestation of the brain in action. When we are conscious, making decisions, getting angry or fearful, these are just functions of the physical brain, nothing more.

A BRIEF HISTORY OF YOUR GREY MATTER

HOW DID WE become the most intelligent creatures on earth? Looking back through the history of our species, we can map out the journey that turned us from simple ape to thinking human. And we can estimate the origins of distinctly human mental abilities, such as when we first began to order our thoughts, when our visual imagination blossomed, when we started to think about the past and future, and when we first played make-believe.

People have long sought a 'secret ingredient' unique to humans that could explain our extraordinary cognitive abilities. Recently, the spotlight has fallen on size – the idea that a big brain is the key. However, it is clear that there is no secret ingredient. Instead, our peculiar way of thinking may have resulted from a reorganisation of the different brain regions, as much as from their expansion.

Tools for thinking

What accelerated this evolution in our ancestral line beyond what was happening in other apes – and how did this give rise to new ways of thinking? Aside from walking on two legs, our early ancestors were distinctly ape-like, and like chimps and other primates they probably had limited tool use, such as picking pebbles off the ground to crush nuts. But things changed around 3.3 million years ago, in what is now Kenya. Rather than just using nature as they found it around them, the ancient hominins began to modify it, wielding one stone to chip the end off another and using the resulting sharp edge to butcher meat.

The idea of using one tool to create a more useful implement is a conceptual leap. But just as important is the fact that it takes dexterity and motor control not seen in other apes to create the tool. This includes coordinating your limbs so that one hand is doing a different job from the other – movements that chimps struggle to master.

Even with that trigger, our ancestors were slow to advance. Things didn't begin to take off until *Homo erectus*, about a million years later. *Homo erectus* is significant for many reasons. As well as having broadly similar bodies to modern humans, they lived in bigger social groups than their predecessors.

Successful communal living requires both cooperation and the ability to detect and punish cheats who try to get something for nothing. Those challenges may have spurred the evolution of complex emotions such as shame and embarrassment, which would help individuals toe the line. But what really marks out the thinking of *Homo erectus* is encapsulated in the genesis of another tool, known as an Acheulean hand axe.

The first attempts at designing such a tool, which date from around 1.5 million years ago, were fairly crude, but over the following million years these leaf-shaped Acheulean axes became thinner and more symmetrical as they began to embody a more systematic style of working.

Breaking down a goal into a series of smaller actions in this way shows the beginnings of hierarchical thinking. Chunking and sequencing our actions seems so central to the way we operate today – whether we are making a cup of tea or running a bath – that it's almost impossible to imagine our minds working in any other way. But the sophisticated later versions of the Acheulean axe offer some of the first signs that our ancestors were beginning to develop the ability to organise their thoughts in these more complicated ways.

Hierarchical thinking has been linked to another milestone in human cognition: language. It is such a complex system, dependent on many different thought processes, that its origins have been described as evolution's biggest mystery, but some evidence suggests that tool-making could have been a catalyst in its development.

Articulate vocalisation requires precise movements of the lips and tongue. Chimps and other primates are unable to achieve these, but for our ancestors tool-making drove the development of the brain areas involved in motor control that were later co-opted for speech. And the sequential thinking needed to create these hand axes is similar to the thinking that allows us to understand and construct sentences.

Origins of speaking

Language is, arguably, humanity's only unique feature, and its emergence set us on a road that led away from every other animal. Unfortunately this turning point in our journey is virtually invisible in the archaeological record. But there are hints that our ancestors had begun speaking by the time of *Homo heidelbergensis*, 600,000 years ago. *Homo heidelbergensis* is thought to have evolved from *Homo erectus* and was certainly more human in some respects. Its brain, at about 1,200 cubic centimetres, was just a shade smaller than ours, providing a cognitive power that is evident in the variety of tools it used, including refined hand axes, cleavers and spearheads. To envisage an amorphous lump of rock transforming into these different shapes and styles would have required good spatial cognition, perhaps signalling the birth of visual imagination.

Homo heidelbergensis also possessed refinements in its vocal anatomy. For instance, traces in bones indicate more nerves linking the brain and tongue than their predecessors, and their voiceboxes seem to lack a balloon-like appendage that constrains vocalisations in other apes. Both of these changes would have been needed to produce eloquent sounds. Whenever it emerged, language brought a whole new set of mental challenges, such as understanding the mental images conjured up by language and a better verbal memory to remember what had been said by others in a group.

Mental leap

Our ancestors were probably still navigating these difficulties as the human mind approached the last stretch of its journey. A demonstration of this mental leap can be found in stone tools made around 300,000 years ago. Unlike the early crude-looking tools from 3.3 million years ago, these so-called Levallois tools would be much more likely to stand out as a work of human ingenuity if you tripped over them on a path. The craftsmanship needed to make them would have taken great skill and patience. They were created by carefully preparing a stone core to form a distinctive turtle-shell shape, then striking it to produce a series of flattish sharp flakes for use as tools such as scrapers, knives and the tips for projectiles. This process comprises many different stages, and needs specific instruction, thus the mind that originally created this tool was probably capable of advanced hierarchical thinking and complex communication. Although these intricate objects are found among the remains of our own species, they are most commonly associated with the Neanderthals.

Levallois tools provide some of the best evidence that Neanderthals shared much of the cognitive toolkit possessed by modern humans living at the same time. And herein lies the mystery. Why did we develop more ambitious inventions and rich artistic cultures, while Neanderthals hit a dead end?

Neanderthal minds

Some people think the answer is child's play – literally. Since our ancestors first diverged from the other primates, childhood has continued to get longer, giving the brain more time to develop outside the womb. From the remains of bones and teeth, it seems that early human children took longer to develop than Neanderthal children.

The extra time spent playing may have helped them develop 'counter-factual thinking' – the ability to consider how things might be, not just how they are. That allowed them to imagine the environment in more creative ways, giving them greater control over their surroundings. As a result, they could do things that might not have occurred to earlier humans, like inventing new tools and building shelters.

Others put this last cognitive leap down to a chance mutation that increased our ability to hold several ideas in mind and manipulate them. Even in modern humans, this 'working memory' is limited to about seven items. However, a small increase would have had huge consequences. An improved ability to remember what had just been said would have increased the sophistication of conversation, allowing more complex grammar with many different clauses. That means you can think and plan more hypothetically, using 'what if' and 'if, then' statements, for instance. Working memory is also associated with creativity and innovation, because it allows you to explore different solutions to a problem in your mind.

Brain food

Further clues come from the food available at the time. Early modern humans began to hunt and trap game, such as small deer species and rodents, which could entail placing snares in ten to fifteen places to capture food. But that requires forethought, and keeping track of the location of the snares – further evidence of a leap in working memory performance.

The timing of these advances at 70,000 years ago is significant because they come just after the eruption of the Toba supervolcano in Indonesia, which plunged the world into a mini ice age that caused a human population crash in Africa. Any beneficial mutations within the small remaining population could therefore spread quickly, leaving a permanent mark on their descendants. This was the beginning of the home stretch to the modern mind.

Armed with this slightly superior thinking we moved over the globe, while the Neanderthals and our other evolutionary cousins became extinct.

THE POWER OF DREAMS

Until about 2 million years ago, human ancestors probably settled for the night in trees. Dozing on branches is likely to have ended with *Homo erectus*, who, at six feet tall and 140 pounds, was too tall and heavy. Sleeping on the floor may have resulted in a great leap in cognition. A more peaceful night's slumber, without the risk of falling from a branch, would have allowed *Homo erectus* to spend longer in rapid eye movement (REM) sleep and slow wave sleep. These stages are crucial for the consolidation of memories and linking different thoughts and ideas, a key process for creative thinking.

YOUR BRAIN THROUGH THE AGES

THROUGHOUT LIFE OUR brains undergo more changes than any other part of the body. These can be broadly divided into five stages, each profoundly affecting our abilities and behaviour.

Setting the stage: gestation

By the time we take our first breath, the brain is already more than eight months old. It starts to develop within four weeks of conception, when one of three layers of cells in the embryo rolls up to form the neural tube. A week later, the top of this tube bends over, creating the basic structure of fore-, mid- and hindbrain.

From this point, brain growth and the development of different regions is controlled mainly by genes. Even so, the key to getting the best out of your brain at this stage is to have the best prenatal environment possible. In the early weeks of development, that means having a mother who is stress-free, eats well and stays away from cigarettes, alcohol and other toxins. Towards the end of the brain-building process, when the foetus becomes able to hear and remember, sounds and sensations also begin to shape the brain.

In the first two trimesters of pregnancy, though, development is all about putting the basic building blocks in place: growing neurons and connections and making sure each section of the brain grows properly and in the right area. This takes energy, and a variety of nutrients in the right quantity at the right time. In fact, if you consider the size of the construction job at hand – 86 billion brain cells and several million support cells in four major lobes and tens of distinct regions, from a starting point of nothing – it is a truly staggering feat of evolutionary engineering.

Soaking it up: childhood

In childhood, the brain is the most energetic and flexible that it will ever be. As we explore the world around us it continues to grow, making and breaking connections at breakneck speed. Perhaps surprisingly, learning, memory and language begin before we are even born. During the prenatal period, up to a quarter of a million new cells form every minute, making 1.8 million new connections per second, though about half of the cells will later wither and die, leaving only those reinforced by use. From birth, a child undergoes more than a decade of rapid growth and development, in which every experience contributes to the person they will become.

Birth alters brain function surprisingly little. Although the touch-sensitive somatosensory cortex is active before birth, it's another two or three months before there is any other activity in the cortex, which ultimately governs such things as voluntary movement, reasoning and perception. The frontal lobes become active at six months and a year old, triggering the development of emotions, attachments, planning, working memory and attention. A sense of self develops as the parietal and frontal lobe circuits become more integrated at around eighteen months, and a sense of other people having their own minds at age three to four.

Life experiences in these early years help shape our emotional well-being, and neglect or harsh parenting may change the brain for good. Maternal rejection or trauma early in life, for example, may affect a person's emotional reactions to stressful events later on, potentially predisposing them to depression and anxiety disorders.

By age six, the brain is 95 per cent of its adult weight and at its peak of energy consumption. Around now, children start to apply logic and trust and to understand their own thought processes. Their brains continue to grow and make and break connections as they experience the world until, after a peak in grey matter volume at eleven in girls and fourteen in boys, puberty kicks in and the brain changes all over again.

Wired, and rewiring: adolescence

Teenagers are selfish, reckless, irrational and irritable, but given the cacophony going on inside the adolescent brain, is it any wonder? In the teenage years our brains may be fully grown, but the wiring is still a work in progress.

Psychologists used to explain the characteristics of adolescence as the product of raging sex hormones; more recently, though, imaging studies have revealed structural changes in the teenage years and early twenties that go towards explaining these tumultuous years. Adolescence brings waves of cerebral pruning, with teenagers losing about 1 per cent of their grey matter every year until their early twenties.

This pruning trims unused neural connections that were over-produced in the growth spurt of childhood, starting with the more basic sensory and motor areas. These mature first, followed by regions involved in language and spatial orientation and lastly those involved in higher processing and executive functions.

Among the last to mature is the dorsolateral prefrontal cortex at the very front of the frontal lobe. This area is involved in control of impulses, judgement and decision-making, which might explain some of the less-than-stellar decisions made by the average teenager. This area also acts to control and process emotional information sent from the amygdala – the fight-or-flight centre of gut reactions – which may account for the mercurial tempers of adolescents.

As grey matter is lost, though, the brain gains white matter. This fatty tissue surrounds neurons, helping to conduct electrical impulses faster and stabilise the neural connections that survived the pruning process.

These changes have both benefits and pitfalls. At this stage of life the brain is still childishly flexible, so we are still sponges for learning. On the other hand, the lack of impulse control may lead to risky behaviours such as drug and alcohol abuse, smoking and unprotected sex.

On the plus side, as teenagers rush towards adulthood and independence, they carry with them the raw potential to sculpt their brains into lean, mean processing machines. Making the most of this time is a matter of throwing all that energy of youth into learning and new experiences. But whether they like it or not, while their decision-making circuits are still forming, their brains still need to be protected, if only from themselves.

The slippery slope: adulthood

So you're in your early twenties and your brain has finally reached adulthood. Enjoy it while it lasts. The peak of your brain's powers comes at around age twenty-two and lasts for just half a decade. From there it's downhill all the way.

This long, slow decline begins at about twenty-seven and runs throughout adulthood, although different abilities decline at different rates. Curiously, the ones that start to go first – those involved with executive control, such as planning and task coordination – are the ones that took the longest to appear during your teens. These abilities are associated with the prefrontal and temporal cortices, which are still maturing well into your early twenties.

Episodic memory, which is involved in recalling events, also declines rapidly, while the brain's processing speed slows down and working memory is able to store less information. So just how fast is the decline? From our mid-twenties we lose up to 1 point per decade on a test called the Mini–Mental State Examination (MMSE). This is a 30-point test of arithmetic, language and basic motor skills that is typically used to assess how fast people with dementia are declining. A 3- to 4-point drop is considered clinically significant. In other words, the decline people typically experience between the ages of twenty-five and sixty-five has real-world consequences.

There is an upside. The abilities that decline in adulthood rely on 'fluid intelligence' – the underlying processing speed of your brain. But so-called 'crystallised intelligence', which is roughly equivalent to wisdom, heads in the other direction. So even as your fluid intelligence sags, along with your face and your bottom, your crystallised intelligence keeps growing along with your waistline. The two appear to cancel each other out, at least until we reach our sixties and seventies.

There's another reason to be cheerful. Staying mentally and physically active, eating a decent diet and avoiding cigarettes, booze and mind-altering drugs seem to slow down the inevitable decline. And if it is too late to live the clean life, don't panic. You still have a chance to turn it around.

Down but not out: old age

By the time you retire, there's no doubt about it, your brain isn't what it used to be. By sixty-five, most people will start to notice the signs: you forget people's names and the teapot occasionally turns up in the fridge.

There is a good reason why our memories start to let us down. At this stage of life we are steadily losing brain cells in critical areas such as the hippocampus – the area where memories are processed. This is not too much of a problem at first; even in old age the brain is flexible enough to compensate. At some point though, the losses start to make themselves felt. Clearly not everyone ages in the same way, so what's the difference between a jolly, intelligent oldie and a forgetful, grumpy person? And can we improve our chances of becoming the former?

Exercise can certainly help. Numerous studies have shown that gentle exercise three times a week can improve concentration and abstract reasoning in older people, perhaps by stimulating the growth of new brain cells. In fact, your brain is doing all it can to ensure a contented retirement. During the escapades of your twenties and thirties and the trials of midlife, it has been quietly learning how to focus on the good things in life. By sixty-five we are much better at maximising the experience of positive emotion.

So while nobody wants to get older, it's not all doom and gloom. In fact you should probably stop worrying altogether. Studies show that people who are more laid-back are less likely to develop dementia than those who are more stressed. In one study, people who were socially inactive but calm had a 50 per cent lower risk of developing dementia compared with those who were isolated and prone to worry. This is likely to be caused by stress-induced high levels of cortisol, which may cause shrinkage in the anterior cingulate cortex, an area linked to Alzheimer's disease and depression in older people.

Our brains may not wrinkle and sag like our skin, but they need just as much care and attention – so don't give up on yours too soon.

LEARNING IN THE WOMB

Experiences during the late prenatal period are certainly important, and perhaps vital for normal brain development. Learning can first be detected experimentally at about twenty-two to twenty-four weeks of gestation, when fetuses will respond to a noise or a touch but will ignore the same stimulus if it occurs repeatedly – a simple kind of memory called habituation. From around thirty weeks fetuses show conditioning – a more complex kind of memory in which an arbitrary stimulus can be learned as a signal that something will happen, like a sound signalling a poke. Fetal memories for particular pieces of music and the mother's voice and smell have all been shown to form sometime after thirty weeks' gestation and to persist after birth.

Language acquisition begins prenatally. A newborn will suck more vigorously if it hears its native language rather than a foreign one. Talking to a third-trimester fetus helps them to recognise your voice, but there is no direct evidence that exposure to multiple languages in the womb will influence future linguistic talents. The most important factors in language development are how much a parent talks to the child after birth, the complexity of their vocabulary and how well they focus the child's attention.

MAPPING THE MIND

EVER WONDERED WHAT'S going on inside your head? Not in terms of thoughts, ideas and memories, but in terms of the hardware. How does the tissue inside your skull actually generate your feelings, emotions and sense of awareness?

If you could open up your skull and pull out your brain, the most striking thing you would notice is that the brain resembles an oversized greyish-pink walnut, covered in deep folds and wrinkles. This outer layer of the brain is the cerebral cortex. Not all animals have brains that look like this. Some, like rats and mice, have very smooth brains, while others, like pigs and people, have ridges and furrows. Mammals with larger brains have a more folded cortex, and the human brain is the most wrinkled of all, cramming as much grey matter into our skulls as possible.

Another striking feature is that the brain is divided into two halves – just like a walnut. The left side controls the right side of the body, and vice versa. When you wave your right hand, it's the left side of your brain at work. This means that damage to one side of your brain affects the opposite side of the body.

You've no doubt heard that the brain's right hemisphere is more creative and emotional, and the left deals with logic, but the reality is more complex. Nonetheless, the sides do have some degree of specialisation. The left side deals with speech and language and processes social cues, and the right deals with spatial and body awareness.

Attempts to understand the brain's architecture began with reports of people with brain damage. One famous example is the case of Phineas Gage, a nineteenth-century railroad worker, who lost part of the front of his brain when a 1-metre-long iron pole was blasted through his left cheek and out of the top of his head. His survival and subsequent change in personality made him one of neuroscience's most famous case studies – one of the first to highlight that specific areas of the brain affect particular aspects of behaviour. We now know that local-ised damage results in highly specific impairments of particular skills – such as literacy or numeracy – suggesting that the brain is modular, with different locations responsible for different mental functions.

Advanced-imaging techniques developed in the late twentieth century gave a more nuanced approach by allowing researchers to peer into healthy brains as volunteers carried out different cognitive tasks. The result is a detailed map of where different skills arise in the brain – an important step on the road to understanding our complex mental lives.

Neuroscientists divide the brain into three main parts, which carry out different types of processing and evolved in different periods of our evolutionary history: the hindbrain, midbrain and forebrain.

Hindbrain

As its name suggests, the hindbrain is located at the back of the skull, just above the neck. It is the most primitive part of the human brain, with its precursor emerging in the earliest vertebrates. When someone talks of their 'lizard brain', this is the part of the organ that they are referring to. This brain region is responsible for many of the automatic behaviours that keep us alive, such as breathing, regulating our heartbeat and swallowing.

The most prominent part of the hindbrain is the cerebellum. Sometimes known as the 'little brain', the cerebellum looks different from the rest of the brain because it consists of much smaller and more compact folds of tissue. It represents about 10 per cent of the brain's total volume but contains 50 per cent of its neurons. The cerebellum's main job is to control voluntary movements and balance, and it is also thought to be involved in our ability to learn specific motor actions and to speak. Problems in the cerebellum can lead to severe mental impairment and movement disorders.

Midbrain

At the base of your brain is the midbrain, a small but significant region that plays a role in many of our physical actions. It regulates eye movement, sleep and arousal, and conveys sensory and motor information to other brain areas.

One key area of the midbrain is the substantia nigra, so-called because it is a rich source of the neurotransmitter dopamine, which turns black in post-mortem tissue. Since dopamine is essential for the control of movement, the substantia nigra is said to 'oil the wheels of motion'. It is the region affected in Parkinson's disease, causing tremors, stiffness and difficulty moving.

Forebrain

By far the largest part of your brain is the forebrain. Many of our uniquely human capabilities arise in this area, which expanded rapidly during the evolution of our mammalian ances-tors. It includes the thalamus, a relay station that directs sensory information to the cerebral cortex for higher processing; the hypothalamus, which releases hormones into the bloodstream for distribution to the rest of the body; the amygdala, which deals with emotion; and the hippocampus, which plays a major role in memory.

The most obvious part of the forebrain is the cerebrum, which accounts for the majority of the brain's mass. It is split into two halves – or hemispheres – divided by a deep groove and covered by the wrinkled layer of cerebral cortex. Here plans are made, words are formed and ideas generated – it is the home of our creative intelligence, imagination and consciousness.

Structurally, the cerebral cortex is a single sheet of tissue made up of six crinkled layers folded inside the skull; if it were spread flat it would stretch over 1.6 square metres. Information enters and leaves the cortex through about a million neurons, but it has more than 10 billion internal connections, meaning the cortex spends most of its time talking to itself.

Each cerebral hemisphere is further subdived into four lobes. At the back is the occipital lobe, devoted to vision, and the parietal lobe above that, dealing with movement, position, orien-tation and calculation. Behind the ears and temples lie the temporal lobes, dealing with sound and speech comprehension and some aspects of memory.

To the fore are the frontal and prefrontal lobes, often consid-ered the most highly developed and most 'human' of regions, dealing with the most complex thought, decision-making, plan-ning, conceptualising, attention control and working memory. They also deal with complex social emotions such as regret, morality and empathy.

Another way to classify the regions is as sensory cortex and motor cortex, controlling incoming information and outgoing behaviour, respectively.

The two cerebral hemispheres communicate with each other via a tract of about a million axons, called the corpus callosum. Cutting this bridge, a procedure sometimes performed to alleviate epileptic seizures, can split the feeling of having one unified sense of 'self'. It is as if the body is controlled by two independently thinking brains. One smoker who had the surgery reported that when he reached for a cigarette with his right hand, his left hand would snatch it and throw it away.

So this is how your brain is put together. But to really understand what is going on, you need to zoom in closer, to the level of the cells that make up your brain.

Inside grey matter

Every thought you have and action you make ultimately boils down to the action of the brain's fundamental building block: the neuron. Vast networks of these tree-like cells shift information around the brain. Our billions of neurons, joined by trillions of neural connections, build the most intricate organ of the body.

We know about these nerve cells thanks to the Spanish anatomist Santiago Ramón y Cajal. While investigating the anatomy of neurons in the nineteenth century, he proposed that signals flow through neurons in one direction. The neuron gathers incoming information from other cells and transmits it along the neuron's nerve fibre, called the axon. These fibres can vary considerably in length: the ones that extend from the base of the spine to the toes can be more than a metre in length.

These messages are transmitted as brief pulses of electricity.

They carry a small voltage – just 0.1 volts – and last only a few thousandths of a second, but they can travel great distances during that time, reaching speeds of 120 metres per second. The nerve impulse's journey comes to an end when it hits a synapse – the gap between nerve cells – triggering the release of molecules called neurotransmitters, which carry the signal to other neurons. These molecules briefly flip electrical switches on the surface of the receiving neuron. This can either excite the neuron into sending its own signal, or it can temporarily inhibit its activity, making it less likely to fire in response to other incoming signals. Each is important for directing the flow of information that ultimately makes up our thoughts and feelings.

Most surprisingly, Ramón y Cajal noted that insect neurons matched and sometimes exceeded the complexity of human brain cells. This suggested that our abilities depend on the way neurons are connected, not on any special features of the cells themselves. Ramón y Cajal's 'connectionist' view opened the door to a new way of thinking about information processing in the brain, and it still dominates today.

You only have to open your eyes to experience this connectivity for yourself. As you gaze around, you are totally unaware of the fragmented nature of the brain's information processing activity going on underneath. All these tasks are combined smoothly: depth, shape, colour and motion all merge into a three-dimensional image of the scene.

It's not just grey matter that matters

We often speak of our 'grey matter', but the brain contains white matter too. The grey matter is the cell bodies of the neurons, while the white matter is the branching network of thread-like tendrils that spread out from the cell bodies to connect to other neurons. White matter contains high concentrations of a substance called myelin, which forms sheaths around the axons of nerve cells. This fatty tissue is like insulation along a cable – allowing electrical impulses to zip along faster. Confusingly, in the living brain these tissues are not grey and white. They get their names from how they look after brain tissue has been removed and prepared in the laboratory.

HUMAN BRAIN VS PRIMATE BRAIN

Human brains, particularly the cerebral hemispheres, are bigger and better-developed than in other primates. But corrected for body size, the differences are surprisingly small. The difference between a human brain and a chimp or gorilla brain appears to be largely the way neurons are connected. Humans have several unique genes that seem to control nerve cell migration as the brain develops, and various patterns of gene expression in the brain. So the machinery looks much the same, but it certainly works differently.

As for non-primates – other mammals have smaller brains, with less well-developed lobes at the front of the brain. Further down the evolutionary tree, animals lose the cortex altogether, with reptiles having a brain that resembles our own brainstem. In simple animals, the brain becomes more of a swelling at the top of the nerve cord or around the mouth area.

02.
PERCEPTION

HOW YOUR BRAIN INVENTS REALITY

YOUR SENSES ARE your windows on the world, and you probably think they do a fair job at capturing an accurate depiction of reality. Don't kid yourself.

The most basic – and arguably most important – function of the brain is to take in information from the outside world and process it to create an internal representation of reality. The study of sensory perception is one of the oldest in neuroscience, yet still one of the most surprising. Many common-sense assumptions turn out to be false. Did you know that you have at least 22 senses? Or that most of the time you're blind and what you 'see' is your brain filling in the gaps? Or that the brain is really a prediction machine that makes guesses about what will happen next? Rather than capturing an accurate screengrab of the world outside your head, perception is largely made up by your brain.

Although it feels as if you are looking out at a continuous widescreen movie, most of the time your eyes are only gathering information from a tiny part of the visual field. At the back of the eye is a small patch of densely packed photoreceptors called the 'fovea', the retina's sweet spot, the only part of the eye capable of seeing with the rich detail and full colour we take for granted. This tiny spot – which covers an area of our visual field no bigger than the moon in the sky – is the source of almost all the brain's raw visual information. The reason we see the bigger picture is that our eyes constantly dart about, fixating for a fraction of a second and then moving on. These jerky movements are called saccades, and we make about three per second, each lasting between 20 and 200 microseconds.

While saccades are happening we are effectively blind. The brain doesn't bother to process information picked up during a saccade because the eyes move too rapidly to capture anything useful. It's a bit like blundering in the dark waving a flickering torch with a narrow beam. Despite the fact that you don't normally notice saccades, you can catch them in action. Looking at your eyes close-up in the mirror, flick your focus back and forth from one pupil to another. However hard you try you cannot see your eyes move – even though somebody watching you can. That's because that motion is a saccade, and your brain isn't paying attention. Now pick two spots in the corners of your visual field and flick your gaze from one to the other and back again. You might notice a brief flash of darkness. This is your visual cortex clocking off.

Frozen time

How your brain weaves such fragmentary information into a seamless movie is a mystery. Perhaps memory retains information from previous fixations that it integrates into the here-and-now. You can get a feel for this from the frozen-time illusion – when you look at a clock and the second hand appears to freeze momentarily before tick-tocking back into action. This happens because, to compensate for the temporary shutdown of vision during the saccade, your brain retrospectively makes a guess at what it would have seen, back-filling the 100 or so milliseconds of blindness with the image that comes after the saccade. If your eyes happen to alight on the clock just after the second hand has moved, your brain assumes that the hand was in that location for the duration of the saccade too. The 'second' then lasts about 10 per cent longer than normal, which is enough for you to notice.

This guesswork isn't confined to vision. The auditory system is also full of gaps and glitches that the brain cleans up as it goes. We encounter situations in which people's voices are obscured or distorted, yet we still understand what is said. This is because of a phenomenon called phonemic restoration.

Alien speech

One demonstration of the brain's ability to extract meaning from distorted sounds is a form of synthesised speech called sine-wave speech. When you first hear a sentence in sine-wave speech it sounds alien and unintelligible, a bit like whistling or birdsong. But if you listen to the same sentence in normal speech first, the sine-wave version suddenly becomes crystal clear. This happens because the brain has circuits that respond to speech, but doesn't switch them on unless it detects spoken language. Sine-wave speech isn't speech-like enough to trigger the circuits, but once you know it is speech they spring into action.

This all adds up to one thing: your internal representation bears little resemblance to actual reality: it is more like a hallucination.

MIND MATHS

One of the brain's biggest challenges is making predictions from its crackling electrical storm of activity – the words likely to crop up next in a conversation, for example, or whether a gap in the traffic is big enough to allow you to cross the road. But what lies behind this crystal-ball gazing?

The answer might be that it runs on a type of mathematics known as Bayesian statistics. Named after Thomas Bayes, an eighteenth-century mathematician, this is a way of calculating how the likelihood of an event changes as new information comes to light. According to the Bayesian brain theory, the brain is a probability machine that constantly makes predictions about the world and then updates them based on information from the senses. This theory has been used to explain how the brain builds up a picture with each sweep of our gaze, correcting any errors as it goes. It can also explain the phantom pains and sounds people experience during sensory deprivation: they come from the neural processes that are at work as the brain casts about wildly to predict future events when there is little information to help guide its forecasts.

YOU ARE HALLUCINATING RIGHT NOW

BIRDIE BOWERS, A member of Captain Scott's ill-fated exped-
ition to the South Pole in 1911, knew something was amiss
when a herd of cattle appeared in the distance of the featureless
Antarctic landscape. In ancient times, a strange experience like
this might have been treated as a message beamed directly
from the gods; more recently hallucinations were recognised
as common symptoms of mental illnesses such as schizophrenia. Now, though, it is clear that they occur in people with
perfectly sound mental health. Around 5 per cent of us will
experience at least one hallucination in our lifetime, and the
likelihood increases when we reach our sixties.

The feeling of being real

Hallucinations are sensations that appear real but are not
elicited by anything in our external environment. They are
not only visual – they can also come in the form of sounds,
smells, even touch. It's difficult to imagine just how real they
seem unless you've experienced one. Musical hallucinations,
for example, are not so much imagining a tune in your head
but more like listening to the radio.

The hallucinations of people of sound mind have led to a

better understanding of how the brain goes about creating a world that doesn't really exist, revealing their important role in our perception of the real world, and perhaps how the very fabric of our reality is constructed. Brain scans of people having visual hallucinations show that brain areas active during the hallucination are also active while viewing a real version of the image. In the brain, hallucinations look a lot like real perception.

Hallucinations in people who have recently lost a sense may reveal the most about how the brain works. In the early 1700s, the Swiss scientist Charles Bonnet described his increasingly blind grandfather beginning to hallucinate the appearance of people in his room, wearing majestic cloaks of red and grey. Something similar can happen with other senses. A person who is losing their hearing might hallucinate music, and someone who has lost their sense of smell might hallucinate strong odours.

Sensory deprivation

Sensory loss doesn't even have to be permanent to bring on hallucinations. When our senses are diminished, all of us have the potential to hallucinate. This is what gave rise to the illusion of a herd of Antarctic cattle for the explorer Bowers. His hallucination happened during a whiteout, when a combination of cloudy sky and featureless snowy surface deprived him of visual information from the landscape, so his mind made something up. Truck drivers on long, empty roads sometimes experience something similar. And when put inside an anechoic chamber, a room that is so silent that you can hear your eyeballs moving, people generally start to hallucinate within twenty minutes of the door closing behind them.

One explanation for this is that the brain doesn't tolerate a lack of external information. When starved of data from the real world it turns inward, where the sensory regions chug away with a low-level of spontaneous activity. Usually this activity is suppressed and corrected by real sensory data coming in from the world, but when starved of information, such as in the deathly silence of an anechoic chamber, the brain turns to its internal churnings to make predictions about what is happening. A second possibility is that in the absence of external input, the brain puts too much emphasis on internally generated sounds. The sound of blood flowing through your ears isn't familiar, so it could be misattributed to coming from outside the body and so set off a train of hallucinogenic thought.

Stream of consciousness

Although bombarded by thousands of sensations every second, the brain rarely stops providing you with a steady stream of consciousness. When you blink, your world doesn't disappear. Nor do you notice the hum of traffic outside your home or office or the tightness of your socks. Processing all of these things all the time would be a very inefficient way to run a brain. Instead, it takes a few shortcuts. Most of the time, the brain uses its predictions to fill in the missing pieces and to keep the world around you in line with expectations.

Interestingly, the sense of touch seems to be the exception to the rule. Your skin is never short of sensory input: from the chair beneath you to the tag in your sweater. To cope with this onslaught of sensory information and still have enough processing power to guard against real threats, your brain has a formula to decide what requires attention. Touches that are rapid – 250 milliseconds apart or less – are dismissed, so away go the chair and the sweater tag. Your tactile brain is in the grip of a constant 'reverse hallucination'. It feels nothing even though there is actually a lot going on.

Whichever way hallucinations operate, they work via a fundamentally similar mechanism: the brain decides what you will see, feel and hear irrespective of what's happening around you.

WHEN HALLUCINATIONS GO WRONG

In schizophrenia, hallucinations can get out of hand. Tests on people with the condition often show overactivity in their brain's sensory cortices, and poor connectivity from these areas to their frontal lobes, which are the areas judging what is most likely to be real. This suggests that their brain makes lots of predictions that are not given a reality check before they pass into conscious awareness.

These insights are helping to provide strategies for treatment. People with drug-resistant schizophrenia can sometimes reduce their symptoms by learning to monitor their thoughts, understand their triggers and reframe their hallucinations in a more positive, and less distressing, light. This seems to give them more control over the influence of their internal world.

THE MIND'S EYE: SORTING OUT WHAT YOU SEE

CAN YOU TELL a snake from a pretzel? We may not give it much thought, but our ability to perceive the world visually is a feat that even the most sophisticated robots cannot match. From a splash of photons falling on the back of our eye we discern complex scenes made up of various objects, some near, some far, some well lit, some shaded, and many of them partly obscured by others. The information from the photons hitting a particular spot on the retina is restricted to their wavelength (colour), and their number (brightness). Turning that data into meaningful mental images is a tough challenge, because so many variables are involved. The number of photons bouncing off an object depends both on the brightness of the light source and on how pale or dark the object is.

It is in the visual cortex, located at the back of the brain, where much of the processing takes place. When items obscure each other, the brain must work out where one thing ends and another begins, and take a stab at their underlying shapes. It must recognise the same objects from different perspectives: consider the image of a chair viewed from the side compared with from above. Then there's the challenge of recognising novel objects – a futuristic new chair, for example.

Filling in the gaps

So how does the brain work its magic? In the early twentieth century, psychologists used simple experiments on people with normal vision to glean some basic rules that they called the 'gestalt principles'. For example, the brain groups two elements in an image together if they look similar, having the same colour, shape or size, for example. And if not all of an object is visible we mentally fill in the gaps.

These principles can only go part of the way to describing visual perception, though. They explain how we separate different objects in a scene, but they cannot tell us how we know what those objects are. How, for instance, do we know that a teacup is a teacup whether we see it from above or from the side, in light or in shadow? We do this so effortlessly that scientists have turned to people who have disorders of the visual system, known as visual agnosias – where the brain struggles to recognise things due to an inability to process sensory information, usually caused by some kind of brain damage – to help make sense of it.

Visual confusion

Our visual system is split into several specialist areas. Rather than memorising each object that we have ever seen, our brains construct objects from a series of smaller building blocks, then mentally map how these parts fit together. One person with a visual agnosia was shown a series of three-dimensional objects, each made from two simple geometric shapes. Afterwards he was shown a stream of these images, with a few new objects thrown in and asked if he had seen each object before. People with normal vision scored close to 100 per cent, but he made some intriguing mistakes. He knew he hadn't seen an object before if it contained a new part, but those that had the same parts in a different configuration confused him.

Other agnosias have revealed how the brain makes sense of what an object can do. Some people with agnosia can use objects, but not consciously describe them, while other people can describe objects but not use them easily. Brain-imaging has shown that these two skills – describing objects and using them – are handled in different parts of the visual cortex. One pathway is necessary to perceive an object (known as the 'perception–action' system), while another (the 'what–where' system) deals with an object's physical location and guides the movement of our bodies. These can be broken down yet further. Brain scans suggest that shape, texture and colour are all processed in individual regions. In fact, the closer neuroscientists look, the more modular our visual systems seem. Yet while our brains process objects as collections of parts and features, we experience them as a seamless whole. When we consciously see something, all these disparate elements are stitched seamlessly together, so we know instantly that an apple is smooth, green and round. The brain achieves this by binding all the different features of an object together, perhaps in a separate part of the brain. Crucially, it is only once this link has been formed that an image can pop into our consciousness.

Evidence for this comes from people with simultanagnosia, a kind of visual agnosia in which people experience the world in an unglued-together way – where the objects in a scene are perceived separately, rather than as a whole. A person with this condition, when looking at their place setting on the dinner table for example, might see just a spoon, with everything else a blur. These people tend to have damage to their brain's posterior parietal lobe, which may be crucial for linking information from the visual pathways and bringing it into our consciousness.

Studying people with these unusual ways of perceiving the world is not only shedding light on how the mind sorts out what we see, but could help unlock one of the most fascinating mysteries of modern neuroscience: how the brain binds together all of our sensory experiences into a single, flowing conscious experience that we call 'the present moment'.

THE MAN WHO MISTOOK HIS WIFE FOR A HAT

'He reached out his hand and took hold of his wife's head, tried to lift it off and put it on. He had apparently mistaken his wife for a hat!' This is neurologist Oliver Sacks's description of one of his most intriguing patients, Dr P, who gave rise to the title of Sacks's famous book about strange neurological disorders. Dr P, a professional musician, had a razor-sharp mind except when it came to certain aspects of vision, due to a tumour or degeneration of the visual parts of his brain. Dr P had lost the ability to see the whole picture, failing to recognise a flower, for instance, which he described as 'a convoluted red form with a linear green attachment'. He couldn't recognise faces either, not even of his family or himself. Yet he saw faces in inanimate objects, mistaking water hydrants and parking meters for children.

MAKING SENSE OF YOUR SENSES

CLOSE YOUR EYES. Now stretch out your arms and wiggle your fingers. How do you know where your arms are? How do you know your fingers are moving? Now do it all again, standing on one leg. If you fell over, did it hurt?

Clearly you have your senses to thank for managing this feat. But probably not the five you learned about at school. The idea that we only have five senses comes from Aristotle – and nearly everyone can name them as sight, hearing, touch, taste and smell. Yet none of these senses tell you where your body is in space, whether it is moving and whether you are feeling pain. So it stands to reason that the body has more than just the basic five to help it make sense of the world.

Drawing the line

There is disagreement about how many senses we have, not least because it isn't clear how we should define a 'sense'. We could classify them by the nature of the stimulus, into chemical (tastes and smells), mechanical (touch and hearing) and light (vision). We could then subdivide these further, considering a sense to be a system that responds to a specific type of signal. For instance, taste could be seen not as one chemical sense but five – sweet, salt, sour, bitter and 'umami', a Japanese word for the taste of glutamate, which gives us our sense of meaty flavours. Vision could be considered to be one sense (light), two (light and colour) or four (light, red, green and blue). Wherever we choose to draw the lines it is clear that sensation alone isn't the key to perception, When we talk about our senses, what we are really describing is our perceptions. And the two aren't necessarily the same thing.

Given how naturally perception comes to us, it's perhaps surprising that you don't need it to survive. Much of the planet's life gets by with just one or two basic senses – typically light and touch. A plant grows to follow the apparent motion of the sun, whereas a Venus flytrap closes in reaction to an insect touching hairs on the inside of the trap.

We, however, do much more than this. We see light and shade but use this to perceive objects, spaces and people, and their positions. We hear sounds, but we perceive voices or music or approaching traffic. We taste and smell a complex mixture of chemical signals, but we perceive the mix as ice cream or an orange or a steak. Perception is the 'added value' that the organised brain gives to raw sensory data. Perception goes way beyond sensations, stirring memory and higher-level processing into the mix, too.

Listening but not hearing

With sound, one kind of processing allows the brain to determine the direction of the noise. We can screen out one sound when attending to another. In the well-known 'cocktail party phenomenon', we ignore all extraneous sounds while taking part in a conversation, but can quickly switch focus if someone else mentions our name. The implication is that we are always 'listening' to ambient sound but not always 'hearing' it, except when it suddenly becomes meaningful. In all sensory domains, perception goes far beyond the bare sensation.

This ability puts us at an evolutionary advantage over animals that only have sensations to play with. A simpler creature might be easily fooled by brightly coloured flowers, or markings that look like eyes, but a highly perceptive animal that can add context is far less at the mercy of its senses.

Even so, perception is not without its flaws. For instance, there is the strange case of synaesthesia, a mixing of the senses.

Sensory mix-up

The most commonly reported forms are experiencing sounds, letters, numbers or words as colours, which results from cross-wiring in the brain areas responsible for processing these sensations and concepts. There are many other types of this bizarre sensory mix-up. One musical synaesthete, for instance, not only sees certain colours when she hears specific notes but can 'taste' certain notes. Some notes taste like mown grass; others are like cream. And in 2008 the first case of touch-emotion synaesthesia was reported, where texture gives rise to strong emotions. The sensation of denim evoked feelings of depression and disgust, in one such synaesthete. Possibly we all have this facility to a greater or lesser extent, which is why minor chords are 'sad', why blues music is 'blue' and food can taste 'sharp'.

The way that we understand our experience begins with the senses but goes further than that to build a rich and constant understanding of the world. It's not so much the sensory information that matters, but the meaning that your brain is able to build from it.

EXTRA SENSES

Human senses only go so far. Other creatures have sensory superpowers.

Feeling the electricity: Fish, some amphibians, duck-billed platypuses and even some types of dolphin can sense the electric fields generated by nearby creatures.

Pimped-up colour vision: Some insects and birds have four, five or even six colour receptors, allowing them to perceive colours that are impossible for us to experience or even imagine. For them, the three-colour world of human vision would be as dull as greyscale.

Magnetic sense: Many species – including pigeons, sea turtles, chickens, naked mole rats and possibly cattle – can detect the earth's geomagnetic field.

Echolocation: Certain species of bat get much of the detail they need to find food through echolocation: clicks, squeals and screams that they belt out at up to 120 decibels in ultra-sound, above the range of human hearing. Some blind humans have learned to echolocate with clicks too.

HOW YOUR BRAIN CREATES 'NOW'

WHAT IS 'NOW'? It is not something 'out there' that we can detect with our senses. So how do we perceive time flowing from the past to the present with a brief interlude that is the present moment?

'Now' is a slippery concept. It is an idea that physics treats as a mere illusion, but we couldn't operate in the world if the present moment had no duration. So how long is now? Neuroscientists and psychologists have an answer. 'Now' – the window within which your brain fuses what you are experiencing into a 'psychological present' – has a duration, and it lasts, on average, between 2 and 3 seconds.

This was demonstrated in a neat experiment. Volunteers watched short movie clips in which segments lasting from milliseconds to several seconds had been cut into chunks and shuffled randomly. If the shuffling occurred within a segment of up to 2.5 seconds, people could still follow the story as if they hadn't noticed the switches. But the volunteers became confused if the shuffled window was longer than this. In other words, our brains seem able to integrate jumbled stimuli into a cohesive, comprehensible whole within a time frame of up to 2.5 seconds. This window is the 'subjective present', and exists to allow us to perceive consciously sequences of events.

This time frame crops up in our lives in all kinds of intriguing ways. Movie shots rarely last less than two or three seconds, unless the director is aiming to create a sense of chaotic or confusing movement. And hugs, kisses and handshakes tend to last around three seconds on average. But these short segments aren't the smallest unit of time that the brain deals with. Each moment is in turn made up of a jumble of subconscious mini-nows, from which only a portion are admitted into consciousness. The length of each mini-now depends on the timescale at which our senses can distinguish one event from another. This varies for different senses. The auditory system, for example, can distinguish two sounds that are just 2 milliseconds apart, whereas the visual system requires tens of milliseconds. Detecting the order of these sensations takes even longer. Two events must be at least 50 milliseconds apart before you can tell which came first.

Feeling the flow

Each 2–3 second segment is then sewn together into the smooth-flowing river of time in which the present moves smoothly into the past. One theory about how this happens is that the brain maintains a hierarchy of nows, each of which forms the building blocks of the next, until the property of flow emerges. This sense of continuity operates over a timespan of about thirty seconds and may be held together by working memory – the ability to retain and use a limited amount of information for a short time.

One mystery about how we measure 'now' comes from the well-known stretchiness of time. If there is such a thing as 'now', why does time sometimes seem to drag and sometimes to fly? There is plenty of anecdotal evidence that time can seem to expand or contract depending on what's happening around us – for example, events seem to unfold in slow motion during car accidents. Such expansion has also been reproduced in the lab: when people are presented with a succession of stimuli of equal length they report that an unexpected event in the series seems to have a longer duration. What's more, experiments suggest that when people perceive an event to have lasted longer than it actually did, they also take in more detail about it, describing it more accurately. This may indicate that temporal stretchiness reflects real changes in sensory processing. It might even have conferred an evolutionary advantage. By ratcheting up the brain's processing rate at critical moments and easing back when the environment becomes predictable and calm again, we conserve precious cognitive resources for when we really need them.

Such changes in sensory processing would be subconscious, but might we be able to take control of our perception of 'now'? Regular meditators often claim that they live more fully or intensely in the present than most people, and some experiments back this up. Meditators were asked to look at a visual illusion that appears to flip between two images, and to press a button each time their perspective of it reversed. The reversal time in this kind of task is considered a good estimate of the length of the psychological present.

Both groups perceived 'now' to last about 4 seconds, but when they were asked to try to hold a given perspective for as long as possible, the meditators managed 8 seconds on average, compared with 6 seconds for the others. This suggests that their training, which may reflect improved attention skills and working memory capacity, allows them to stretch time in the moment.

Perhaps with a bit of effort we are all capable of manipulating our perception of 'now'. If meditation extends your 'now', then as well as expanding your mind it could also expand your life. So grab hold of your consciousness and revel in the moment for longer. There's no time like the present.

CALENDAR SYNAESTHESIA

Some people experience the passage of time completely differently to the rest of us. They actually 'see' time: not as a vague conceptual timeline, but as a vivid calendar that feels so real they could almost touch it. One person with this little-known variation of synaesthesia, in which the brain links one kind of sensation to another, sees time as a hula hoop that anchors 31 December to her chest and projects the rest of the year in a circle that extends about a metre in front of her. For another calendar synaesthete the year is a backwards C hovering before her, with January at one end of the horseshoe and December at the other. When she thinks of a date she feels herself travel along the calendar to the right spot. These people seem to experience a supercharged version of the way everyone else experiences time.

MAGICOLOGY: MAGIC AND THE BRAIN

WE SHAKE OUR heads in disbelief as coins are conjured out of thin air, as cards are mysteriously summoned from a pack, and as the magician's assistant vanishes before our eyes. Of course, there is no such thing as 'magic', so how are magicians' tricks so convincing?

After years of ignoring these masters of sleight of hand, neuroscientists and psychologists now realise that the methods magicians use to manipulate the human mind might hold important insights. We know that while magic tricks appear to break the laws of nature, they don't really. So the key to magic must lie in the human brain, packed as it is with glitches and weaknesses ripe for exploitation.

Not paying attention

A good starting point for exploring the art of magic is the magicians' own classification of their trade into three broad types of trick: misdirection, illusion and forcing. Misdirection lies at the heart of magic. It is a way of diverting the audience's attention away from the act of deception and towards something the magician chooses. In neuroscience terms, misdirection relies on the fact that the brain has a very limited supply of attention.

Focusing on one thing can make you oblivious to other things that would otherwise be obvious. This bizarre phenomenon is called inattentional blindness, and it was famously demonstrated in 1999 by psychologists at the University of Illinois at Urbana-Champaign. They made a video of six people in a circle passing two basketballs around. When asked to count the number of passes, around half of the people who watch the video fail to notice someone in a gorilla suit walking through the middle of the game and beating their chest (see Pay Attention!, page 109).

Magicians use this 'inattentional blindness' to pull off blatant deceptions right under our noses. When something seems to 'disappear', for example, it may have simply dropped out of view while your spotlight of attention is directed elsewhere.

At other times, magicians want to hold your attention rather than divert it. Recent studies compared two different kinds of hand movements used by a professional magician in different kinds of tricks. It turned out that slow, circular hand motions are good at engaging and holding our attention, while fast, straight ones are useful for quickly diverting it from one spot to another. Another way of manipulating attention is with humour. Some magicians use jokes to conceal large movements that are particularly difficult to hide. Exactly why laughter disengages attention so effectively is not well understood.

Seeing what's not there

Then there are illusions, which rely on the fact that much of what you think you see is actually invented by your brain. In the vanishing ball illusion a magician tosses a small ball up and down while following it with his eyes. He fakes a third toss, keeping the ball in his hand but still moving his eyes as if watching it. This reliably creates the illusion of the ball being thrown upwards – then disappearing into thin air.

Tracking people's eyes as they watched this trick revealed that on real throws, the eye movement of subjects followed the ball's trajectory. But on the trick toss, their eyes remain firmly glued on the eyes of the magician. This shows that the brain overrules the eyes and creates an image of an object that doesn't actually exist. Why would it do that? Part of the answer lies in the power of social cues – in this case the magician's eyes – to set up expectations in the brain. The trick works less well if you keep your eyes fixed on the throwing hand rather than tracking the arc of the non-existent ball. The trick also relies on a glitch in visual perception. Information captured by the retina takes about 100 milliseconds to reach the brain. To compensate for this lag, the brain predicts what the world will look like in the near future and acts on this prediction rather than the real information at its disposal. This is useful in real-world situations such as driving a car, but it also gives magicians an opening that they can exploit.

Illusion of free will

A third tool up the magicians' sleeve is forcing. This is any technique that gives the target the illusion of free will when in fact they have none. The classic example is the 'pick a card, any card' trick where the magician uncannily seems to know what you picked.

Great magicians, through countless hours of practice, manipulate our attention, memory and causal inferences, using a bewildering combination of visual, auditory and tactile methods. The greatest magic show on earth, though, is the one continually happening in your brain.

MADE UP MEMORIES

Your memory may feel like a reliable record of the past, but it is not. We readily form strong 'memories' of events that have never happened to us. Each time we remember something we, in effect, rewrite history. Magicians learned about this brain glitch long before neuroscientists. During a trick, a magician will often describe what they have just done in a way that manipulates people's recollection of it. So while it might feel like the magician read your mind, actually, they wrote in something new.

TRICK YOUR SENSE
OF TOUCH

IMAGINE YOU ARE lying in the bath with your toes poking out of the water. A drip starts to form on the tap, then drops onto your big toe. Ooh! Not pleasant – but was the drip boiling hot or icy cold? It's impossible to tell.

What you just experienced was a tactile illusion – and it's something psychologists are increasingly interested in. Some tactile illusions have been known for a long time but in general they have been harder to explore than visual illusions. However, now that researchers have started to develop new ways to probe the sense of touch tactile illusions are enjoying a golden age.

One reason for the interest is the drive to add tactile or 'haptic' interfaces to phones and other consumer devices. When you switch your phone to vibrate or play a video game with a rumbling controller (which vibrates at certain points in the game, such as when you fire a gun) you are using haptic technology. The plan is to go beyond those elementary applications: for example, adding interfaces so that you can feel who is calling without taking your phone out of your pocket, or an MP3 player you can search by touch alone.

Tactile illusions can be harder to experience than visual ones, but there are many that can be achieved with a little bit of care, perseverance and a few ordinary household items.

The Aristotle illusion

Cross your fingers, then touch the end of your nose. It feels like you are touching two noses. This is an example of what is called 'perceptual disjunction'. It arises because your brain has failed to take into account that you have crossed your fingers. Because your nose touches the outside of both fingers at the same time – something that rarely happens – your brain interprets it as two separate objects. There's also the reverse Aristotle illusion: cross your fingers and touch the inside of a corner of a room or a box. This time, because the wall is contacting the insides of your fingertips, you should feel one surface, not two. Some people even experience three.

A similar effect can be achieved by holding your hands in front of you, palms down. Close your eyes and get somebody to lightly tap the back of both hands, one after the other, as quickly as possible. Open your eyes and wave the hand that was tapped first. You'll get it right every time. Now do it again with crossed arms. If the taps are sufficiently close together – less than 300 milliseconds or so – you'll get it wrong a lot of the time.

This stems from a failure to 'remap' your body schema to take your crossed hands into account, but that can't be the whole story as single taps are easy to get right even with crossed hands. Neuroscientists think it happens because your brain is trying to remap your body schema and work out the order of the taps at the same time. The second task interferes with the remapping and causes it to fail.

Boxing clever

Take two cardboard boxes of different sizes and put a brick in each one. Check that they weigh the same, then get somebody to lift them and tell you which is the heavier. The vast majority of people will say that the smaller box is heavier, even though it isn't, and will continue to insist that it is even after looking inside both boxes and lifting them several times. This 'perceptual size-weight illusion' is very robust. So much so that it works even if the smaller box is slightly lighter. Even labelling two identical boxes 'heavy' and 'light' can pull the same trick.

The exact reason for these illusions remains a mystery. Curiously, experiments show that even though people initially use greater force to lift the larger box than the smaller one, on subsequent lifts they unconsciously equalise the amount of force they use to lift them. Despite their bodies apparently 'knowing' that the boxes weigh the same, their minds still perceive the smaller box as being heavier.

I feel it in my fingers

Your fingertips are among the most sensitive parts of your body, and this makes them surprisingly easy to fool. Take an ordinary comb and pencil and lay your index finger along the top of the comb, then run the pencil back and forth along the side of the teeth. Even though the teeth are moving from side to side in a wave-like motion, your finger will feel as if a raised dot is travelling up and down the comb. This works because the unfamiliar motion of the teeth causes similar skin deformation to the more usual action of running your finger over a raised bump, so your brain interprets it that way.

Your tongue can be fooled in a similar way. Take a fork and press the tip of your tongue between the prongs. It will feel as though the middle two prongs are bent out of shape. This is because the skin on your tongue is distorted in a way that doesn't normally happen, so your brain assumes that the prongs, not your tongue, are bent.

PLUG-IN SENSES

The fact that our sense of touch is so easily fooled makes it easy to hijack for other purposes. In 1969, neuroscientist Paul Bach-y-Rita rigged up a television camera to a chair, on which was a 20-by-20 array of stimulators that translated images into a pattern of vibrations on the person's back. Blind participants quickly learned to use it to detect horizontal, vertical and diagonal lines, and with practice to 'see' faces and common objects. He later developed a smaller version that translated the camera's images onto a postage-stamp-sized array of electrodes on the tongue.

Other researchers are working on a similar idea, using tactile sensors built into a buzzing vest. In theory, it could be programmed to turn sensory information that we can't naturally detect, from ultraviolet light to infrasound, into something that the brain can understand.

A SPECIAL CASE
FOR FACES

HAVING SPENT A perfectly nice evening on a date with a woman, Jacob Hodes spent the rest of the year at college ignoring her. He didn't set out to play it cool. He simply could not remember what his date looked like. He had had the same trouble all his life: people would say 'Hi', and he wouldn't have a clue who they were. He knew the names of the latest celebrities but could have walked past any one of them in the street without noticing them.

This all made sense when he was diagnosed with prosopagnosia, or face blindness, a condition that means he is unable to recognise people by their faces. It is a little-known condition, but is so common that if you're not prosopagnosic yourself, you almost certainly know someone who is. It seems to run in families but, strange as it might sound, until they're tested most people who have it don't even realise.

Friends and strangers

Research into prosopagnosia is not only about helping people improve their dating skills. It might also explain how the rest of us recognise faces. We are a highly social species, and being able to tell friends from strangers that might harm us is a key skill that is likely to have been favoured through evolution. Some of our face-recognition skills seem to come hardwired as part of our basic brain package. Babies prefer to look at faces over anything else from the moment they are born, which suggests that while we are great at perceiving the world in general, faces are special.

Prosopagnosia was first recognised in 1947 when neurologist Joachim Bodamer described the condition in a twenty-four-year-old man who'd lost the ability to recognise his friends and family, and even his own reflection, after a bullet wound to the head. Until fairly recently only a few dozen cases had ever been described, all caused by brain injury, and the condition was considered extremely rare.

Alternative strategies

Then came the discovery of a second form of face blindness, 'developmental prosopagnosia', which is either present from birth or develops early in life. This affects up to 2 per cent of people. Strangely, it seems that if you have never known what it is to recognise a face, you don't necessarily know that you're supposed to be able to do it. Many of those with developmental prosopagnosia recognise people by the way they walk, or their clothes or voice. Most developmental prosopagnosics are so good at using these strategies that, unless they see a familiar person out of context, with a new hairstyle or in different clothes to normal, they can recognise people well enough.

The brain signature of prosopagnosia hasn't quite been nailed down yet, but we do have some important clues. A key area for face recognition is the fusiform gyrus, also called the fusiform face area. Neuroscientists have known for many years that this area lights up in response to faces, and also to patterns suggestive of a face.

There is almost certainly more to face-processing than one brain area, however. Recent brain-scanning experiments revealed that the brains of those with prosopagnosia were wired slightly differently from most people. In those without prosopagnosia, regions towards the front of the brain form a 'hub', which is highly connected to other regions, including those at the back that process more basic visual information. But people with prosopagnosia were missing this hub-like behaviour. They also had a greater number of connections in the back portions of the brain, which might be an attempt to compensate for what they were missing in terms of connections from the front. Not everyone with prosopagnosia has the same degree of face blindness – those who were better at face recognition showed a greater number of connections in this region. The bigger their deficit in face-processing, the greater the hyper-connectivity.

Super-recognisers

At the other end of the scale there are super-recognisers – people with incredibly accurate powers of face recognition. They were totally unknown to science until a handful of people who had read about research into face blindness contacted researchers to say that they had the opposite condition – that they never forgot a face. All said they sometimes hid their exceptional ability for fear of making people uncomfortable, or to avoid coming across as a stalker. When they were tested in the lab, these people did so well on face recognition tests that a new, more difficult test had to be devised. When given the new test they still scored much higher than average.

London's Metropolitan Police has identified several super-recognisers among its officers, who use their skills to make a disproportionate number of arrests from CCTV footage of crimes. Psychologists are studying these super-cops to see if they can glean any clues about the way we perceive faces.

It isn't yet clear how or whether a normal face recogniser can become super, but there is some evidence that children with prosopagnosia can improve with training, and the younger they start the better.

A SENSE OF PLACE

Our brains are remarkably good at helping us make sense of space. Specialised cells in the brain's hippocampus and nearby regions make a mental map of our surroundings and compute our place within it. Not everyone, however, has this luxury. People with a condition called developmental topographical disorientation (DTD) struggle to form mental maps and so can't easily orientate themselves. This means they get lost – a lot.

But practice can give navigation skills a boost. The hippocampus region in the brains of London taxi drivers, for instance, seems to increase in volume in response to the navigational challenge of their job. This has spurred researchers to look into ways to help people with DTD. They have come up with one idea that might help us all: play more computer games. One study showed that playing *Super Mario* every day helped averagely bad navigators improve their mental map-making skills.

03.
INTELLIGENCE

WHAT EXACTLY IS INTELLIGENCE?

INTELLIGENCE IS A bit like obscenity: hard to define, but you know it when you see it. Intelligent people are, well, brainy. They are good at pretty much anything that requires mental effort: mathematics, verbal reasoning and the like. They learn more easily, grasp concepts more quickly and are better at solving problems. And their big brains tend to bring success in life. On average, intelligent people live longer, healthier lives and earn more money. No wonder we value intelligence highly.

Despite this – or possibly because of it – the study of intelligence is among the most controversial areas of science. Its history is littered with disreputable and pseudoscientific ideas such as phrenology and racial hierarchies. Today, intelligence remains contentious, not least because there is still no scientific consensus on what the word means, or how to measure it. Nonetheless, it is self-evident that, whatever intelligence is, some people have more of it than others.

Intelligence spectrum

The modern scientific study of intelligence began in the late nineteenth century with the work of Francis Galton, Charles Darwin's cousin. He combined the observation that people's mental abilities varied with Darwin's work on biological variation, and proposed that intelligence was distributed a bit like height. In other words, most people cluster around the mid-point, with ever-fewer people towards both ends of the spectrum. He also suggested that intelligence was measurable, and tried to do it (rather unsuccessfully) using reaction times.

The challenge of measuring intelligence was picked up in the early 1900s by the psychologist Charles Spearman. He noted that schoolchildren who performed well in one subject tended to perform well in other apparently unrelated ones. So, for example, a high score on a maths test predicts a high score on one of vocabulary, and vice versa. Spearman proposed that this reflected some deeper, general ability and called this the general factor of intelligence or g.

Cognitive complexity

More recent work on g has shown that it correlates with diverse features of the brain, including processing speed and relative size. But while the existence of g is now generally accepted, it is still difficult to pin down biologically. A high IQ is not something you can spot from a brain scan. Differences in g probably reflect differences in the brain's information-processing efficiency. At the behavioural level, g is proficiency at mentally manipulating information, which underpins learning, reasoning, and spotting and solving problems. In essence, g equates to an individual's ability to deal with cognitive complexity.

From a user's point of view, however, g is not much use. It is not an individualised 'score' that quantifies your intelligence, but a statistical construct derived from the aggregate performance of many people. To put a number on your own general intelligence, you have to turn to yet another controversial measure, IQ – of which more later.

Spearman's ideas did not go uncontested, with some critics arguing that *g* failed to capture types of cognitive ability that are not so readily tested in exams, such as a lawyer's verbal dexterity and an engineer's spatial skills. The most vocal critic was psychologist Louis Thurstone, who argued for seven 'primary abilities', including verbal comprehension and spatial visualisation. They eventually reached a compromise. Thurstone conceded that all his primary abilities depended largely on *g*, while Spearman came to accept that there are multiple subsidiary abilities in addition to *g*.

This one-plus-many resolution has been further developed into the 'three stratum theory'. At the top is the single universal ability, *g*. Below are lesser abilities, all related to *g* but each also containing a different 'additive' that boosts performance in one of eight broad domains. These include processing speed (the time it takes to do a mental task) and general memory skills. The third stratum is composed of sixty-four even narrower abilities, including the ability to discriminate sounds in speech, learned skills such as reading comprehension and highly specialised aptitudes such as spatial scanning – the ability to visualise a path out of a maze. This arrangement of abilities is complex but captures the many differences in individuals without undermining the dominance of *g*. For example, an engineer might have excellent visuospatial perception as well as a high *g* score.

The dominance of *g* is also acknowledged by IQ tests, which provide an easy-to-understand measure of general intelligence. The tests are designed so that the average IQ score is 100, with 90 per cent of individuals scoring between IQ 75 and 125. Somebody scoring 130 or above is considered highly intelligent.

IQ controversy

Search for 'IQ test' online and you will be bombarded with options. The gold standard are orally administered, one-on-one tests such as the Stanford-Binet and Wechsler, which require little or no reading or writing and take up to 90 minutes to complete. They test domains such as comprehension, vocabulary and reasoning and combine them to give an overall IQ.

IQ tests remain controversial with some scientists arguing that they don't measure anything meaningful or are culturally biased. However, the very best are the most technically sophisticated of all psychological tests and undergo the most extensive quality checks. If you want to measure your intelligence, whatever that might be, they are your best option.

WHY ARE WE INTELLIGENT?

We have always struggled to explain how intelligence evolved. One theory is that social groups, with their constantly shifting allegiances, selected more intelligent animals. Animals that live in bigger groups tend to have bigger brains, for example. Additionally, intelligence may be favoured when food is scarce. Black-capped chickadees that are descended from birds that lived in demanding environments are better at solving problems than neighbours from more comfortable environments. They also produce more new brain cells and have better memories. The trouble is, we don't know if intelligent animals live longer than dim ones, or produce more offspring.

CAN YOU BOOST
YOUR IQ?

INTELLIGENCE HAS ALWAYS been tricky to quantify, not least because it seems to involve large tracts of the brain and so is almost certainly not one 'thing'. Even so, scores across different kinds of IQ tests show that people who do particularly well – or badly – on one kind of test seem to do similarly on all. This can be crunched into the single general intelligence factor, g, which correlates quite well with intelligence tending to lead to academic success, a higher income, and better health.

So more intelligence is clearly a good thing. Can we get more of it?

First, the bad news. Broadly speaking, your intelligence is fixed by the genes you happen to have inherited. Studies of large numbers of twins show that the IQ scores of identical twins raised apart are more similar to each other than those of non-identical twins raised together. In other words, shared genes have more influence than shared environment, suggesting that nature is more important than nurture.

That doesn't mean the environment plays no part, at least in childhood. While the brain is developing, everything from diet to education and stimulation plays a huge part in developing the structures needed for intelligent thought. Children with a bad diet and poor education may never fulfil their genetic potential.

But even for well-fed and educated children, the effects of environment wear off over time. In young children, genes account for 30 per cent of the variation in intelligence scores. By young adulthood, that has risen to 60 to 80 per cent. Adult identical twins raised apart score almost exactly the same on IQ tests, while adoptees in the same household may as well be strangers. The conclusion is that most family environments are equally effective for nurturing intelligence – your adult IQ will be the same almost regardless of where you grew up, unless the environment is particularly inhumane.

Maximum cognitive ability

So if genes play such a big part, is there anything adults can do to improve IQ? Not much, as it turns out, though not nothing either. It is worth remembering that IQ tests are designed to measure maximum cognitive ability, but we rarely perform at that level in everyday life. Lack of sleep, stress, hunger, illness and hangovers all blunt our cognitive tools, including intelligence. So the best way to max your IQ is to avoid all of these.

Brainpower also needs protecting over the long term. Chronic illness, alcohol abuse, smoking and head injuries cause cumulative damage, but such assaults are preventable.

There are also active ways of slowing or reversing losses in cognitive function. The most effective discovered so far is physical exercise, which shields the brain by protecting cardiovascular health. Healthy body, healthy mind may be a cliché, but that is because it's true. Mental exercise, often called brain-training, is sometimes touted as a way to increase IQ, but it only boosts the particular skill that you practise. What is more, there is little evidence that the thing you get better at is transferrable outside the brain-training programme itself.

Quick fix

Drugs offer another quick fix. Caffeine and nicotine both increase alertness for short periods, and may help you to perform close to your maximum IQ. Pharmaceuticals designed for other things are also widely used as 'smart drugs'.

Surveys of university students often report extensive use of the drugs Ritalin and Adderall to boost memory and concentration; both are used to treat attention-deficit hyperactivity disorder (ADHD). Another favourite is modafinil, designed to treat narcolepsy and other sleep disorders but which can also reduce fatigue and maintain alertness – useful when you need to pull an all-nighter. However, such drugs do not increase intelligence, they only enhance certain aspects of cognition such as memory or alertness. Many people are unwilling to take the risk of side effects or legal sanctions.

That might help explain the rise of so-called superfoods as natural brain boosters. Unfortunately, while eating blueberries, salmon, avocados and dark chocolate is considered safer, it may not be as effective. If such 'brain foods' work at all, it is probably by promoting general health.

Some people even try transcranial direct current stimulation (tDCS), which involves placing electrodes on the scalp to pass a weak electrical current through the brain. The equipment is cheap and safe, and has been claimed to improve specific functions such as working memory, mental arithmetic, focused attention and creativity. But these are controversial claims, and if there is an effect it is only transient.

The good news is that one type of intelligence keeps on

improving throughout life. Most researchers distinguish between fluid intelligence, which measures the ability to reason, learn and spot patterns, and crystallised intelligence, the sum of all our knowledge and experience gained throughout life. Fluid intelligence slows down with age, but crystallised intelligence keeps on rising. So while we all get a little slower to the party as we get older, we can rest assured that we are still getting cleverer, almost without trying.

ARE SOME BRAINS WIRED BETTER?

At Einstein's autopsy in 1955, his brain was something of an anticlimax: it turned out to be a little smaller than average. Indeed, studies have suggested a minimal link between brain size and intelligence. It seems brain quality rather than quantity is key.

One important factor seems to be how well our neurons can talk to each other. The brains of more intelligent people seem to have more efficient networks between neurons – in other words, it takes fewer steps to relay a message between different regions of the brain. That could explain about a third of the variation in a population's IQ. There are indications that Einstein's brain had particularly efficient neural networks.

Another key factor is myelin, the insulating fatty sheath encasing neuron fibres. Better insulation means that nerve impulses travel faster, and there is a significant correlation between the quality of myelination and IQ.

ARE WE GETTING MORE INTELLIGENT?

IN DENMARK, EVERY man is liable for military service at the age of eighteen. Nowadays only a few thousand get conscripted but all have to be assessed, and that includes measuring their IQ. Around 30,000 young men are tested every year. Until recently, the same test had been used since the 1950s.

Look back over those decades of IQ data and a clear trend emerges. Year after year, the average IQ kept on nudging upwards, to around 3 extra points per decade. That may not sound a lot but the cumulative rise is such that what would have been an average score in the 1950s is now low enough to disqualify a man from military service.

The same phenomenon of ever-rising IQ scores has been observed in many other countries. For at least a century, each generation has been slightly but measurably brighter than the last. As a result IQ test scores have to be 're-normalised' every decade or so, to ensure that the average score stays at 100.

This steady rise in test scores has come to be known as the 'Flynn effect' after James Flynn of the University of Otago in New Zealand, who was one of the first to document it. There's no question that the effect is real and that intelligence – as measured by IQ tests, at least – has risen. In the United States, for example, average IQ rose by 3 points per decade from 1932 to 1978. Everywhere psychologists have looked, they have seen the same thing. So why have IQ scores been increasing around the world?

Flynn himself is sure his eponymous effect is meaningful, but does not believe that it means we're all becoming geniuses, or that our grandparents were dunces. More likely we have simply developed the skills and habits of mind that make us better at solving the sort of abstract problems that appear on IQ tests. That may be a result of our tech-soaked culture with ubiquitous computers, mobile devices, video games and so on.

If so, the Flynn effect may be less a general increase in intelligence than a turbo boost in certain specialist cognitive skills. An IQ test comprises a series of subtests, and it turns out that scores in some of these have increased a lot, including visuospatial skills and the ability to identify similarities between common objects. Others have not increased at all, such as vocabulary and arithmetic abilities.

Stunted abilities

Another probable contributory factor is that people are no longer held back by poor health and poor environments. It is well known that a bad start in life can stunt physical growth. Something very similar may happen with intelligence. Children whose mothers experienced famine in the Netherlands during the 1940s, for example, had reduced cognitive function, and malnutrition in the early years of life does seem to lead to a lower IQ.

Wherever social conditions start to improve, the Flynn effect kicks in. In post-war Japan, for example, IQ shot up by 7.7 points per decade; two decades later it started climbing at a similar rate in rapidly industrialising South Korea. With improved nutrition, better education and more stimulating childhoods, many people around the world really have become more intelligent. On top of that, better public health measures reduce the need for our immune systems to expend resources to combat infectious disease, leaving us able to spend more on growth – and larger, cleverer brains may be one consequence.

Not only that, as more people travelled and married outside their local group, populations may have benefited genetically from hybridisation. Inbreeding is known to lower intelligence, and outbreeding can raise it.

If better nutrition, public health and education have led to rising IQs, the gains should be especially large at the lower end of the range, among the children of those with the fewest advantages in their lives. Sure enough, that's what testers usually see. In Denmark, for example, test scores of the brightest individuals hardly budged – the score needed to be in the top 10 per cent of the population is still about what it was in the 1950s. The movement has all been at the bottom end.

Tapering off

This all begs the question of whether the Flynn effect can continue indefinitely. The answer, unfortunately, appears to be no. Indeed there is evidence that it has already stopped. In Denmark, the most rapid rises in IQ occurred from the 1950s to the 1980s. Something similar seems to be happening in a few other developed countries, too, including the UK and Australia.

Why does this positive trend now seem to be coming to an end? Perhaps it was only to be expected. If the Flynn effect is a consequence of social improvements, then as factors like education and good nutrition become the norm their intelligence-boosting effects should taper off. Flynn himself has been predicting for some time that the effect will run out of steam. Similarly, increases in height are also tapering off.

But IQ scores are not just levelling out: they appear to have gone into reverse. The first small decline was seen in Norway in 2004 and has since been spotted in Australia, Denmark, the UK, Sweden, the Netherlands and Finland. What is going on, and should we be worried?

It is possible that the decline is simply random, or may reflect a regression in social conditions, such as declining income. That is not a positive development, but is at least reversible. The other possibility is that humans really are getting less intelligent – perhaps because developments in medicine and technology mean that we are no longer under evolutionary pressure to maintain our brains. The Flynn effect may have merely delayed our downfall. But look on the bright side: it probably makes you among the most intelligent human beings who have ever lived.

COGNITIVE OVERLOAD

The increasing complexity of the modern world may be making us more intelligent but, ironically, also more dumb. The essence of intelligence is the ability to deal with cognitive complexity; as the world becomes technologically advanced and interconnected, the amount of cognitive complexity we encounter goes up. This may lead to what is called 'cognitive overload' – a familiar feeling to anyone who has tried and failed to upgrade to a new phone, computer or software package.

EMOTIONAL INTELLIGENCE

WE ALL KNOW someone who is frighteningly intelligent when it comes to facts, but hopeless when it comes to mastering social situations. Their IQ may be stratospheric, but not their EQ – their emotional intelligence.

This concept surfaced in 1990 when psychologists were casting around for a pithy way to sum up qualities such as empathy, self-awareness and emotional control. For a while, the term languished in academic obscurity. Then in 1995 it became the title of Daniel Goleman's bestseller *Emotional Intelligence: Why it Can Matter More Than IQ*. Soon the phrase was popping up everywhere, tantalising us with the idea that we each have an EQ to match our IQ – and that our ability to monitor our own and others' feelings could be a better indicator of future success than the kind of abstract intelligence measured by IQ. There was also the promise that, unlike IQ, our EQ could be improved to better equip us for life's trials.

Social cognition

The underlying idea is that the cognitive skills measured in IQ tests are part of a broader spectrum of information-processing capacities of our brains. After all, we know that humans have particularly sophisticated skills to help us understand each other and specialised brain circuitry devoted to the navigation of social situations.

But how can such a complex thing as emotional intelligence be measured in a meaningful way? Some researchers doubt that skills such as understanding and managing emotions can be reduced to a numerical yardstick in the same way as IQ. Others, however, have devised tests to tease apart different aspects of our emotional cognition. These entail asking people to identify the emotions expressed in, say, a photograph of a frowning face, and by answering questions such as:

George was sad, and an hour later he felt guilty. What happened in between?

A: George accompanied a neighbour to a medical appointment to help out the neighbour.

B: George lacked the energy to call his mother, and missed calling her on her birthday.

Those with a high EQ will realise that scenario B is most likely to account for George's guilty feelings. Tests like this were the first steps in putting emotional intelligence on a scientific footing. Research now shows that those with a high EQ have more rewarding friendships and successful work relationships. Perhaps it's not so surprising that it's easier to get along with other people and coexist in a work environment if you have a good handle on your emotions, and those of others. Many companies now assess the emotional intelligence of prospective employees.

In 2010, the parts of the brain vital for emotional intelligence were discovered, through studies of veterans of the Vietnam War who had injuries to their prefrontal cortex – the region of the brain that plays a crucial role in human social and emotional behaviour. Depending on the specific site of their injuries the veterans were assessed as being poor either at 'experiential' emotional intelligence (the capacity to judge emotions in other people) or 'strategic' emotional intelligence (the ability to plan appropriate responses to social situations). Damage to these regions didn't affect cognitive intelligence, suggesting that emotional and general problem-solving tasks are handled independently in the brain.

Schools of thought

The realisation that high emotional intelligence might lead to a more fulfilling life has also led to a different approach to education. In the UK, for example, the Social and Emotional Aspects of Learning programme was rolled out in 2005, with the aim of improving the emotional intelligence of pupils. In American schools, special programmes teach empathy skills and impulse control.

It might seem obvious that this kind of approach to education would produce a more rounded pupil than a system dedicated only to the learning of facts, or that a workplace full of emotionally intelligent people would be a nicer, more co-operative place, but the research doesn't always concur. In fact some studies reveal a dark side to emotional intelligence, where high-EQ people use their skill for personal gain, for example, strategically disguising their own emotions and manipulating those of others.

A high emotional intelligence can even be a hindrance in

the workplace, according to an analysis of all studies linking emotional intelligence and job performance. This found that a high EQ was a bonus for those working in people-focused jobs such as sales or counselling. If your role involves making a customer or patient feel good, then a talent for understanding which emotions are appropriate for the context is an asset. However, for roles where emotional awareness is not so crucial – scientific research, accountancy or working as a mechanic, for example – high emotional intelligence can lead to lower job performance. The idea is that the effort a person with high EQ puts into 'emotional labour' detracts from the job at hand. Emotional intelligence is not always an asset.

MACHINES THAT CAN READ EMOTIONS

If you're in any doubt of the vast skill it takes to decode human emotions, then consider the challenge to develop computers with this kind of intelligence. Computer programmers have created software that can easily beat a chess grandmaster, but battle to teach computers how to understand human foibles. Things that we do effortlessly, such as sensing a person's emotional state from their facial expression, are fiendishly difficult for a machine to grasp.

This capacity might be crucial for machines of the future. There is a growing recognition that human emotions are essential to decision-making, planning and judgement. So the question facing computer scientists is whether machines without emotions and feelings will ever be able to think and plan effectively.

WHY YOUR HIGH IQ DOESN'T MAKE YOU A GENIUS

SO YOU'VE GOT a high IQ. Congratulations. You'll probably live longer, have more academic success and be better paid than someone with a lower score. But don't get too carried away. High intelligence is a useful tool, but not a guarantee of success. It won't stop you from acting stupidly. It has little connection with emotional well-being or happiness. And even if your IQ is stratospheric, you'll need a lot of other skills to be a genius.

IQ tests are very good at measuring certain mental faculties such as logic, abstract reasoning and how much information you can hold in mind. But the tests fall down when it comes to measuring those abilities crucial to making good judgements in real-life situations. That's because they are unable to assess things such as a person's ability to weigh up information, or whether an individual can override the intuitive cognitive biases that can lead us astray from making sound decisions.

This is the kind of rational thinking we do every day, whether deciding which foods to eat, where to invest money, or how to deal with a difficult client at work. And yet IQ tests – still the predominant measure of people's cognitive abilities – do not effectively tap into it. They fall short of the full panoply of skills that would come under the rubric of 'good thinking'. For these reasons, it is perfectly possible for you to be highly intelligent and at the same time very stupid (just think of the brilliant physicists who insist that climate change is a hoax). Indeed, a survey of members of Mensa (the high IQ society) in Canada in the mid-1980s found that 44 per cent of them believed in astrology and 56 per cent believed in aliens.

Not only are there many more dimensions to human thinking than IQ measures, but IQ scores can be easily skewed by factors such as dyslexia, education and culture. We would all probably fail an intelligence test devised by an eighteenth-century Native American. What's more, a person's IQ score tells us nothing about their conscientiousness and enthusiasm, which are big factors in whether someone fulfils their natural potential. To use an analogy, a high IQ is like height in a basketball player. It is very important, but there's a lot more to being a good basketball player than being tall – just as there's a lot more to being a good thinker, or a genius, than having a high IQ.

Predicting genius

So what does it take to be a genius? It turns out that no measure of intelligence, whether IQ or other metrics, reliably predicts that a person will develop extraordinary ability. In other words, the IQs of the great would not predict their level of accomplishments, nor would their accomplishments predict their IQs. Studies of chess masters and highly successful artists, scientists and musicians usually find their IQs to be above average, typically in the 115 to 130 range, where some 14 per cent of the population reside – impressive enough, but hardly as rarefied as their achievements and abilities.

The converse – that high IQ does not ensure greatness –

holds as well. This was shown in a study of adult graduates of New York City's Hunter College Elementary School, where an admission criterion was an IQ of at least 130 (achieved by a little over 1 per cent of the general population) and the mean IQ was 157 – 'genius' territory by any scaling of IQ scores, and a level reached by perhaps 1 in 5,000 people. Though the Hunter graduates were successful and reasonably content with their lives, they had not reached the heights of accomplishment, either individually or as a group, that their IQs might have suggested. There were no superstars, no Pulitzer or Nobel prize-winners. In fact, most talented adults – be they scientists or entrepreneurs – were never identified as gifted as children.

Born or made?

So if high intelligence doesn't create geniuses, what does? The ability we're so fond of calling talent arises not from innate gifts alone, but from an interplay of fair (but not necessarily extraordinary) natural ability, quality instruction, and a mountain of work. Examine closely even some extreme examples – Mozart, Newton, Einstein, Stravinsky – and you find that their mastery was hard-won.

Exactly how much hard graft is necessary to reach the top is a matter of debate. In the 1990s, studies of musicians showed that the difference between elite performers and their less exceptional peers was not down to any innate differences in talent, but in the amount of deliberate effort the musicians put in to improve their performance. By the age of twenty, for example, top violinists had notched up more 10,000 hours of training on their violins, far more than their less accomplished peers. This '10,000-hour rule' was popularised by the writer Malcolm Gladwell in his book *Outliers*.

Recent research, however, questions this view. It found that only around 20 per cent of high performance in sport and music could be attributed to extra hours of practice, and only 4 per cent for educational achievement. What explains the rest? The researchers suggest that it's down to a combination of innate talent, general intelligence, the age at which a person seriously started to engage in an activity, and their working memory – the ability to keep information in the forefront of their mind.

The debate rages on. But the message is clear. Don't get fooled into thinking your IQ alone will take you to the top.

THE IDIOCRACY HYPOTHESIS

If intelligence is so crucial to humans, why haven't we all evolved to be intellectual geniuses? One theory is that human civilisation eased the challenges driving the evolution of our brains. The idea is that, in the distant past, people whose mutations had slowed their intellect would not have survived to pass on their genes. As human societies became more collaborative, slower thinkers were able to piggyback on the success of those with higher intellect. This theory is often called the 'idiocracy' hypothesis, after the eponymous film, which imagines a future in which the social safety net has created an intellectual wasteland. Although it has some supporters, the evidence is shaky, and we can't easily estimate the intelligence of our distant ancestors.

04.

CONSCIOUSNESS

THE ENIGMA OF CONSCIOUSNESS

ASK YOURSELF THIS: do you feel conscious? The fact that you are even able to consider the question suggests that the answer is probably yes. Now look into the eyes of the nearest human being. Are they conscious too? This time it's much more difficult to be sure. It doesn't matter whether you are gazing into the eyes of your beloved or a complete stranger, there is no way of truly knowing whether they are conscious, and even if they are, whether their experience of consciousness is anything like yours.

These basic problems with understanding consciousness have had philosophers scratching their heads for centuries. In the seventeenth century, René Descartes proclaimed that the body and conscious mind are cut from very different cloth: the body and the brain are made of matter in the same way as other physical objects such as tables, rocks and plants, while the mind is immaterial, and cannot be seen, touched or directly observed. This has set the tone for much of the debate about consciousness ever since.

In 1995, philosopher David Chalmers at New York University updated Descartes' point of view, calling our efforts to understand consciousness 'the hard problem', as opposed to the 'easy problem' of understanding the brain. We can tell, for example, that the brain is made up of a kilogram or so of highly connected nerve cells, some of which are specialised for certain functions. We can also tell that the currency of communication between nerve cells is both electrical and chemical.

But while that all helps us to explain how our eyes inform our brains about the wavelength of light that relates to, say, the colour red, this doesn't tell you anything about what it is *like* to see red. Nor does it give you tools to describe your experience of red to anyone else. Or, as the philosopher Thomas Nagel put it in the 1970s: you could know every detail of the physical workings of a bat's brain, but still not know what it is like to be a bat.

This 'what it's like' aspect of consciousness is difficult to evaluate; our very real, and very personal, experiences of the world are called *qualia* and they have so far proved tricky to explain. Not all mental states have this special 'feel' associated with them – remembering where you put your keys or that it's your birthday tomorrow, for example. The 'what it's like' aspect of consciousness is almost exclusively associated with our experience of bodily sensations. The experience of the taste of a peach or the redness of a flower wouldn't be the same without this strange extra dimension.

More problematic still is the fact that it is impossible to get inside the head of another person to experience their qualia – or even to tell if they have them. It is possible that everyone else is a 'zombie'. Not in the horror movie sense of the word – a philosophical zombie is someone who is totally devoid of qualia but can still react to the world as if they were conscious. Stick this zombie with a pin and it will say 'ouch' and recoil. But that's just a reflex – it feels no pain. No one has yet found a way to tell for sure that the people around us aren't zombies.

The not-so-hard problem

On the other hand, not everyone is convinced that there is a 'hard problem' to solve. Those with a 'materialist' viewpoint believe that we'll be able to understand consciousness – and perhaps find a way to measure qualia – just as soon as we understand enough about the way the brain works.

According to this view, consciousness is a direct result of the workings of the brain – a side effect that emerges as the brain is going about its day job. The leading explanation for this is that the brain acts as a kind of hypothesis-making machine, constantly combining information from all over the brain to construct a current understanding of what is going on in the world. Since the incoming information is constantly changing, the brain needs to keep on top of what is happening now, updating these drafts of the world as it goes along. The resulting consciousness, then, isn't some mysterious out-of-body experience, but a by-product of the flow of information in the body and brain. So, the richness of our experiences and the feeling that there is an 'I' who is experiencing them may be just the brain's best guess about what is going on and how we should react for the best outcome.

While this view is growing in popularity, it doesn't explain all of the mysteries of consciousness. Whether you think of consciousness as a mystical force or a brain-based illusion, it still doesn't explain why we experience qualia in the way that we do, or how they form. Nevertheless, when studying consciousness, the materialist approach brings two advantages. First, there is no need to explain strange interactions between the material brain and the immaterial mind, because in the materialist view, the latter does not exist – it's simply the feeling inside us generated as the brain goes about its everyday business of making sense of the world. And second, it makes the 'hard problem' of understanding consciousness disappear in favour of a drive to explain how the brain accomplishes this trickery. Over the past two decades, this has brought the problem into the realms of neuroscience – it's still a mystery, but at least we have a better idea about where to start looking for answers.

THE PHYSICS OF BEING CONSCIOUS

The complexities of consciousness have not yet been cracked by biologists. Could physics have the answer? Pioneering quantum physicist Erwin Schrödinger entered the fray in his 1944 book *What is Life?*, musing on how our conscious mind controls the actions of atoms in our body. More recently physicist Max Tegmark has argued that consciousness is actually a state of matter, alongside the states of solid, liquid and gas. His argument is that you are simply the product of the rearrangement of some of the food you've eaten in your lifetime. This shows that your consciousness isn't simply due to the atoms you ate, but depends on the complex patterns into which these atoms are arranged.

WHAT'S THE POINT OF BEING CONSCIOUS?

THE AWESOME FEELING of gazing at the night sky. The intense experience of the colours of trees in autumn. Consciousness provides us with a particularly hi-res, vivid window on the world. One that undoubtedly enriches our lives. Yet it's hard to fathom the point of those feelings, in terms of evolutionary benefits. Why did consciousness evolve in the first place?

Evolution tends to favour traits that have some benefit: providing a survival advantage to those lucky enough to have them. Language, for example, helped us work together and share ideas. But defining the benefits of consciousness is trickier, especially when we are not really sure what it is.

We have, however, made some progress in explaining consciousness in terms of brain activity, and this might provide some clues about its purpose. One of the leading theories, the global neuronal workspace model, says that sensory stimuli, such as sights and sounds, are initially processed separately in the brain at an unconscious level. If something crops up that requires further attention – say, a sudden intruder – the sensory information gets broadcast more widely across the brain. At this point it bursts into consciousness. One idea, then, is that the job of consciousness is to manage complex mental tasks by allowing chunks of information to be processed separately for most of the time, and then combined when necessary.

Another possibility is that consciousness arose out of the need to understand other people's minds as our ape-like ancestors began to live in larger social groups. In a bigger group there would be strong evolutionary pressure to keep track of relationships and to understand and pre-empt the actions of others. This might have led to the ability to think about what another person is thinking, and as a side-effect to reflect on our own thoughts. The survival advantage of this would be a greater ability to put more than one person's ideas together for the good of the group.

Origins

Consciousness could have evolved for many reasons – or perhaps none. It's also possible that it doesn't bring any survival advantage at all, and instead is an 'epiphenomenon' that emerges as a side effect of our complex brains. That can feel like a cop-out, so what if we were to take a different tack? Rather than looking for a reason for human consciousness to exist, we can track consciousness to its evolutionary origins and see what advantages it might bring less complex creatures than ourselves.

This approach, too, has its drawbacks. Consciousness leaves nothing to see in the fossil record, so we have to infer its evolutionary history by comparing animals alive today that show signs of consciousness and working back to what their common ancestor might have been able to do.

Some animals make this easier than others. Chimpanzees recognise themselves in the mirror – a skill often taken as a sure sign of consciousness. Scrub jays will sneak back and re-cache food if they see that another bird had watched them hide it – unless the watcher is their mate. When we look at these creatures, we can infer some sort of awareness of self and of others, that looks a lot like what we recognise as consciousness in ourselves.

There's reason to consider a broader benchmark, though. While these signs of higher consciousness are impressive, not every conscious experience is anything like as complex. When we experience colours, odours, aches and pains, they have more to do with perception and emotion than the higher realms of complex thought. Much of our conscious experience involves shades of feeling – objects are comforting or scary, sounds are pleasing or annoying, our body feels good or bad. These kinds of evaluations guide us towards rewards and away from harm – both of which are key survival skills.

If we take this as the start of consciousness, then far more animals make the grade. Mammals, birds and reptiles all show signs of emotional responses, such as an increased heart rate and elevated body temperature when handled, while fish and amphibians do not. This might put the dawn of consciousness at 300 million years ago in the land-based common ancestor of modern reptiles, birds, and mammals. By an even broader measure, even insects get to join the conscious club. An animal is conscious, some researchers propose, if it experiences the world subjectively on its own terms. This allows flexibility in behaviour that goes beyond mere reflexes and captures the distinctive 'me, here, now' element of our own experience.

Stay focused

One measure of subjective experience is selective attention – the ability to focus on just a few elements among all the sensory information available. In experiments, an ability to home in on salient information has been found in insects, vertebrates and octopuses. We know that the common ancestor of these three groups was a very simple organism that resembled a flatworm. Modern flatworms show few, if any, signs of rudimentary consciousness, so it seems a safe bet that the common ancestor also lacked consciousness. If so, consciousness may have evolved separately in the three groups.

As for why, all of the explanations have one thing in common – consciousness occurs in nimble, fast-moving animals that encounter rapidly changing conditions as they move. This points to a need for flexible decision-making as the drive to form our beautiful, conscious minds.

DON'T BE SO SMUG

There's no doubt that human consciousness is special. Whether it is unique in some way or simply richer than that of other animals is still up for debate. Yet many people who study consciousness do not see it as an all-or-nothing quality: while other animals may not have the highly developed and special form of consciousness that we have, some species probably have a glimmer of it. And those animals we think of as most likely to be conscious – apes and dolphins, for instance – are also innovative problem-solvers and tool-makers.

But perhaps we shouldn't feel too pleased with ourselves. Compared with unconscious processing, consciousness is slow and energy intensive, and can only do one thing at a time. And plenty of creatures manage to surf life's ups and downs pretty successfully without it.

MEDITATE TO EXPAND CONSCIOUSNESS

MYSTICS MIGHT TELL you that meditation expands the mind. Science is beginning to show they are right. It seems that meditation really does increase our ability to tap into the hidden recesses of our brains that are usually outside the reach of conscious awareness.

Tapping into subliminal messages

Meditation boosts our ability to pick up on subliminal messages – ones that we see but are unable to recall consciously. This was tested by flashing up the answer SPRING to a question such as 'Name one of the four seasons' for a subliminal 16 milliseconds. Volunteers who had meditated for 20 minutes beforehand were more likely to give an answer that matched the subliminal word, compared with a control group who hadn't meditated. It seems that meditators can access more of what the brain has paid attention to.

Beware the dark side

Meditation has many measurable cognitive, emotional and health benefits – but also some worrying side effects. A small proportion of people attending meditation retreats report panic attacks, terror, hallucinations and even psychotic breakdown. And the training of Japanese soldiers used to include meditation techniques to encourage the loss of sense of self, so that the soldier 'became' the order he received.

Awareness of unconscious thoughts

When you decide to carry out an action such as pressing a button, activity occurs in the brain region that controls the movement before you feel you made the decision. This suggests that the unconscious brain 'decides' when to press the button. In tests, meditators were able to pick up on this unconscious brain activity quicker than non-meditators.

Focused attention meditation

The aim of this form of meditation is to achieve voluntary control of attention and cognitive processes.

Start by sitting with your eyes closed. Then concentrate your mind on your chosen object – say your breathing. Try to keep it there. Probably your mind will quickly wander away, to an itch on your leg, perhaps, or to thoughts of what you will be doing later. Keep bringing it back to the breath. In time this will train the mind in three essential skills: to watch out for distractions, to 'let go' of them once the mind has wandered, and to re-engage with the object of meditation. With practice, you should find it becomes increasingly easy to stay focused.

Open monitoring meditation

This type of meditation aims to gain an awareness of everything that comes into your moment-by-moment experience – thoughts, emotions, bodily sensations – without reacting to it. Instead of grasping at whatever comes to mind, which is what the majority of us do most of the time, the idea is to maintain a detached awareness. Those who develop this skill find it easier to manage emotions in day-to-day life.

HOW THE BRAIN CREATES CONSCIOUSNESS

CONSCIOUSNESS FEELS LIKE an on–off phenomenon: either you're awake and experiencing the world, or you're not. But finding the button – or perhaps the dimmer switch – in the brain that allows us to move from one state to the other has proven difficult. A key question is whether there is a single module in the brain, a 'seat of consciousness' that is responsible for awareness, or whether it emerges as a result of more complicated activity across a number of brain regions.

We do know there are certain brain regions that, when damaged or electrically stimulated, will result in loss of consciousness. The claustrum – a thin, sheet-like structure buried deep inside the brain – is one of them. In 2014, researchers who were doing exploratory surgery to locate the source of a woman's epileptic seizures stimulated an electrode that had been placed near her claustrum. When the team zapped the area, the woman stared blankly into space and did not respond to instructions. As soon as the stimulation ended, she regained consciousness with no memory of the event.

There are other potential switches, including the thalamus, a relay-centre located in the middle of the brain. Many people in a vegetative state, who have no signs of awareness, have damage to the thalamus, or to the connections between the thalamus and the prefrontal cortex, a region at the front of the brain generally responsible for high-level complex thought.

While these regions are clearly important, most of the leading theories of consciousness don't consider a single on–off switch as the most important determinant of consciousness. Instead, there is a focus on the way that information is integrated across the brain, albeit with certain regions acting as integration hubs.

In the global neuronal workspace model, for example, the brain's specialist processing areas constantly churn through sensory information, almost entirely below the radar of consciousness. Only when a number of different brain regions synchronise their electrical activity does one particular part of our experience burst into consciousness. Brain-imaging research backs this up: at the moment when people report having seen an image, their brain activity stabilises for hundreds of milliseconds, almost as if it is pausing to read out the information. When the same experiments were carried out with people who had varying levels of consciousness – from patients who had recently recovered from a coma to those in a minimally conscious state or a persistent vegetative state – the amount of stability in their brain's response to a noise matched the degree to which they were conscious.

Seeing a red triangle

Then there's information integration theory, which ignores the brain's anatomy, and simply says that consciousness is the result of putting information together in such a way that the whole contains more information than the sum of its parts. It is based on the observation that when we become consciously aware of something, we experience it as a unified whole. When you see a red triangle, the brain does not register it as a colourless triangle plus a shapeless patch of red. And when you contemplate a bunch of flowers, it's impossible to be conscious of the flowers' colour independently of their fragrance. The brain brings the sensory data together to produce an integrated experience.

If this theory is correct, loss of consciousness is due to a breakdown in communication across the brain. This is backed up by a study that scanned people's brains as they were slowly anaesthetised. It found that complete unconsciousness was marked by a failure of the cortex to talk to the rest of the brain. On the other hand, brain scans of people who have taken psychedelic drugs such as LSD, ketamine and psilocybin have found that these drugs seem to increase integration across the brain. This raises the intriguing possibility that they could be experiencing a higher state of consciousness.

Interconnectedness

Scientists have hit upon a way of showing the brain's interconnectedness. By (painlessly) stimulating one part of the brain with an electromagnet, a wave of activity propagates across the brain that can be measured via electrodes on the scalp. The pulse acts like striking a bell and neurons across the entire brain continue to 'ring' in a specific wave pattern, depending on the amount of connectivity between individual brain cells.

This method has been used to compare people in a vegetative or minimally conscious state, emerging from a coma, or healthy and awake. It turns out that neurons 'shake' in a distinctive pattern in response to the electromagnetic pulse, depending on a person's state of consciousness. This technique could be used to distinguish locked-in syndrome, where a person is conscious but unable to communicate, from persistent vegetative state, where a person is totally unconscious.

All things considered, it seems unlikely consciousness has a single on–off switch. Instead, its seat seems to be spread through the networked structure of the brain and the connections that link it all together.

BABY STEPS TO CONSCIOUSNESS

In adults, the moment we become consciously aware of something in our environment is linked to a two-stage pattern of brain activity. The first comes when the sensory areas of the brain register a stimulus. Then, about 300 milliseconds later other areas light up, including the prefrontal cortex, which deals with higher-level cognition. Conscious awareness kicks in only after the second stage of neural activity reaches a specific threshold.

Brain-imaging studies show that babies have a similar pattern of activity, just at a slower pace. In twelve-month-old and fifteen-month-old babies the second stage of activity arrived 800 to 900 milliseconds after the stimulus. In five-month-old infants there was a delay of more than one second. So it could be that babies experience the world more or less the same as we do. It just takes them a little longer to get there.

DIY HALLUCINATIONS

YOU DON'T NEED to be delirious or use mind-altering drugs to experience sensory weirdness. There are a number of simple tricks to fool your brain into seeing or feeling things that aren't actually there.

Closed-eye hallucination

This is a simple one. Just shut your eyes and apply gentle pressure to your eyelids. A luminous pattern or 'phosphene' should appear in your field of view. Why? Pressure from the fingers stimulates the retina at the back of the eyes, which fires signals to the visual part of your brain. This causes you to 'see stars'. Some people see phosphenes when they sneeze or get up too quickly.

Ganzfeld effect

You're going to look rather strange doing this, but you will experience the strange effects of sensory deprivation. To experience the ganzfeld effect, play gentle white noise through headphones and cover your eyes with something semi-transparent – a sheet of paper or half a ping pong ball – so that unstructured but uniform light reaches your eyes. Then find a comfortable place to lie down for half an hour (alternatively, you can fork out to spend an hour in a floatation tank). After a while, you may experience all sorts of visual or auditory weirdness, from simple geometric shapes or random noises to more complex phenomena such as the voices or images of people who aren't there. These hallucinations are thought to arise as the brain amplifies 'noise' from sensory neurons as they attempt to detect signals from the outside world. This noise is interpreted by the higher parts of the sensory cortex, which creates the hallucinations.

Waterfall illusion

This illusion has bemused people since Aristotle described it 2000 years ago. If you fix your eyes on a waterfall for a short time, then look at the bank beside it, the bank will appear to drift upwards. To experience this 'motion after-effect' at home, keep your eyes fixed on the centre of a spinning spiral, then straight away look at the back of your hand – which will appear to squirm.

This illusion is caused by neurons tuned to opposite directions of motion. While watching a waterfall or spinning spiral, the brain cells that detect the direction of motion become tired. When the eyes look away, the cells that detect motion in the opposite direction are more active and make a stationary object appear to be moving.

Feel a forcefield around your body

This is a twist on the well-known rubber-hand illusion where an experimenter uses a paintbrush to stroke a volunteer's hand (which is hidden from view) and an adjacent, visible rubber hand. The stroking is done simultaneously at the same speed and place on both the real and rubber hand. Within minutes, most people report feeling the touch of the brushstrokes on the rubber hand as if it belonged to them. But a twist is needed to experience a 'forcefield': the brush doesn't touch the rubber hand, but brushes above it, at the same time as brushstrokes touch the real hand (this means you feel the touch on your real hand but watch the brush move in mid-air, say, about 10 centimetres above the rubber hand). Most people report feeling a 'magnetic force' or 'forcefield' between the paintbrush and the rubber hand below – as if the brush was encountering an invisible barrier – as well as a sense of ownership of the rubber hand. This happens because our brains are aware not just of our bodies but also the immediate space around us.

PAY ATTENTION!

IMAGINE THAT YOU are walking down the street and a passer-by asks you for directions. As you talk to him, two workmen rudely barge between you carrying a door. Then something weird happens: in the brief moment that the passer-by is behind the door he switches places with one of the workmen. You are left giving directions to a different person who is taller, wearing different clothes and has a different voice. Do you think you would notice?

Of course you would, right? Wrong. When researchers at Harvard University played this trick on fifteen unsuspecting people, eight of them failed to spot the change. What this demonstrates is a phenomenon called 'change blindness'. It happens because of a chronic shortage of a crucial mental resource: attention. You are blithely unaware of most of what is going on around you, to the point that you can fail to notice 'obvious' changes in your surroundings.

Almost every useful feature of your brain begins with attention. It determines what you are conscious of at any given moment, and so controlling it is just about the most important thing the brain can do. But as the switching-workmen example demonstrates, our attention systems are easily fooled, and our natural distractibility often leads us to jump to conclusions that aren't necessarily true.

Blind to change

Scientists studying attention spend a lot of time on change blindness because it provides direct access to the attentional system. In the door experiment, the subjects fail to see the change because their attention is elsewhere and the door conceals what would otherwise be attention-grabbing motion. The trick foxes what is known as the bottom-up attention system. This snaps our focus to anything that stimulates the senses: a movement, a loud noise, an email notification or someone tapping you on the shoulder. It's an ancient skill that evolved for a reason – it's no good focusing well enough to knap the perfect spear tip if you get eaten by a lion before you can use it. This system is fast, unconscious and always on (at least, when you are awake). Ignoring such disturbances is physically impossible, so the only way to stop them from hijacking an otherwise productive day is to shut them out: eliminate unpredictable noise, turn off email notifications, disconnect Wi-Fi.

The other, top-down, system is deliberate and focused on a particular goal. It zooms in on the task at hand, and hopefully stays there long enough to get the job done. There is a constant tug of war between the brain networks that control goal-orientated thinking and those that control impulses, which means that top-down attention is prone to losing focus or being rudely interrupted.

The good news is that, sneaky psychology experiments aside, we can tweak our attention settings to stay focused more easily. As well as cutting down on bottom-up distractions by turning off email notifications, putting your phone on silent and so on, some research suggests that it might be an idea to give your brain more to do. According to studies of distractibility, better control of top-down attention comes not by reducing the number of inputs, but by increasing them. In this view, called 'load theory', once the brain reaches its limit of sensory processing, it can't take anything else in, including distractions.

Engaging the senses might mean adding visual aspects to a task that make it more attention-grabbing without making it more difficult – putting a colourful border around a blank document and making the bit you are working on purple, perhaps – or choosing somewhere with a bit of background noise. This strategy seems to work for both distractions and mind-wandering.

There are also signs that cognitive training might help. Researchers working with people with attention-deficit hyperactivity disorder (ADHD) and brain injuries have found that cognitive training, combined with non-invasive magnetic brain stimulation, can improve focus on a task that needs sustained attention. Wider studies are under way, and initial results hint that the right kind of brain-training could help more or less anyone.

Chill out

While we wait, the next best option is learning to chill out in exactly the right way. The parts of the brain associated with attention have been shown to be thicker in the brains of long-term meditators, while other studies have found that attention test scores improved after a short course of meditation. Learning to focus better may be as simple as making time to sit still and focus on nothing more complicated than breathing.

There may be a downside to an unerring focus, however. If you are deliberately concentrating on something, it can render you oblivious to other events that you would normally have no trouble noticing. This 'inattentional blindness' is probably the reason why motorists sometimes collide with objects that they simply 'didn't see'.

One classic example of inattentional blindness was reported in the paper *Did You See the Unicycling Clown?* This showed that three-quarters of people walking along using their phone failed to spot a unicycling clown they passed, compared with just half of people not using their phone. Sometimes you may not believe your brain.

WHEN ZONING OUT IS GOOD

Since attention is so important to the brain's functioning, mind-wandering used to be thought of as a universally bad thing. Now, though, psychologists are realising that it has several upsides.

When we need to plan for an uncertain future, for example, mental meandering can be the perfect tool. Daydreaming has also been shown to be crucial in boosting creativity and problem-solving, by allowing the brain to forge connections between pieces of information we don't link up when we are too focused.

Above all else, daydreaming is part of what makes us human, allowing us to escape the present and move mentally into a consciousness of our own making. And that is far too much fun to give up.

05.

THE UNCONSCIOUS

MEET THE UNSUNG HERO OF YOUR MIND

HUMANS ARE RATHER proud of their powers of conscious thought – and rightly so. But there is one aspect of our cognitive prowess that rarely gets the credit it deserves: a silent thinking partner that whirrs away in the background. Behold the power of the unconscious mind!

Modern notions of a powerful 'subconscious' were invented by Sigmund Freud as part of his theory of psychoanalysis. 'The conscious mind may be compared to a fountain playing in the sun and falling back into the great subterranean pool of subconscious from which it rises', he wrote. For him, the unconscious/subconscious (he used the words interchangeably at first) played an outsized role in governing human behaviour.

Today, neurology and cognitive science have replaced psychoanalytical notions with evidence-based ideas. They show that a major proportion of your thoughts and actions – even things you believe you are in conscious control of – do indeed take place in your unconscious.

Most of the time you are essentially flying on autopilot. The information we perceive in our consciousness is not created by conscious thought. One idea is that the conscious mind steps in only after unconscious processing has taken place and a decision or action has to be made. Our conscious mind carries out this job then fools itself that it was in charge all along, like a lazy boss who claims all the credit for their team's hard work.

The unconscious is not easy to study, not least because it is difficult to analyse mental processes that are outside conscious awareness. And there isn't a neat brain region that we can point to on a brain scan as the location of the unconscious. Some researchers have even used Ouija boards to try to communicate with people's unconscious – studying the small, almost undetectable muscle movements that can move the board's pointer as if by magic.

You can try a similar phenomenon, using what is known as the 'ideomotor effect'. Make a pendulum out of a paper clip and a piece of thread and dangle it over a cross drawn on a piece of paper. Ask yourself a simple yes/no question, such as 'Am I at home?' or 'Do I have a cat?', and tell yourself that if the pendulum swings clockwise, the answer is yes, while anticlockwise means no. Spookily, the pendulum will generally start rotating in the direction of the correct answer.

It looks supernatural, but it's not. The reason it works is that as soon as you ask the question your unconscious brain fires up motor preparation circuits in anticipation of the answer it expects to see. These circuits initiate subtle muscle movements you are not normally aware of – except when they are amplified by a pendulum (or dowsing stick or Ouija board). This is your unconscious brain in action.

A more orthodox technique is subliminal messaging, in which an image is flashed in front of the eyes then quickly replaced with another before the first image can consciously register. The first image is said to be 'masked'. In this way it has been demonstrated that information shown to the unconscious can spill over into conscious thoughts and decisions. For instance, people shown the masked word 'salt' are then more likely to select a related word from a list, like 'pepper'.

Make your mind up time

If these ideas are disconcerting, there may be an upside. Our ability to process information unconsciously may help us to make decisions. In one study people were asked to choose an apartment by one of three methods. These were: making an instant decision, mulling over all the pros and cons for a few minutes, or thinking about an unrelated problem in order to distract them from consciously thinking about the apartments. People chose the objectively best apartment when they used the distraction method. This might be because they were unconsciously mulling over the decision while their consciousness was elsewhere.

Some of these findings have recently been questioned, because others have been unable to replicate them. Yet there is certainly growing attention paid to the powers of the unconscious. Some researchers believe that unconscious deliberation can also explain those 'a-ha!' moments when the answer to a problem seems to come from nowhere, as well as times when a searched-for word comes to mind only after we stop trying.

A different aspect of your mental underworld is reflected in your 'implicit assumptions'. Your subconscious mind isn't just planning and executing actions, it also spends a great deal of time analysing the world, looking for patterns and relationships that can help you navigate through life. The conclusions it comes to are called implicit assumptions – subtle, often unconscious, biases about people and events. For example, if you hear a description of an eminent professor of physics, it's impossible not to make assumptions about this person (many people automatically conjure up an image of an older, white male).

Because we are not in control of our implicit assumptions, and are seldom aware of them, it is possible to develop unconscious prejudices that your conscious mind might find unappealing or even abhorrent – such as associating men with science and women with the arts, or preferring thin people to fat people. We may find ourselves acting on these without having any idea why.

But even with these downsides, there is one thing that we can all agree on: there is far more to our non-conscious thought processes than we once realised.

EMOTIONAL QUEUE-JUMPING

Most of the time, there is a delay of around 50 milliseconds between our visual system registering a word on a screen and that word being consciously perceived. But when attention-grabbing emotional words such as 'love' or 'fear' are flashed onto the screen, they break through into consciousness a few milliseconds earlier.

This 'fast track' system may help us to act more speedily when we see something that is particularly important to us. It also raises the possibility that it is the unconscious that makes the decision about whether our surroundings warrant conscious inspection.

WHAT YOUR MIND GETS UP TO WHEN YOUR BACK IS TURNED

LIKE IT OR not, your life is ruled by thoughts you have no control over. Our brain has an uncanny knack for working things out with no need for conscious involvement. Even when you are fast asleep, it carries on making sense of the world. It even plans your future.

Say you want to wake up at 6 a.m. Some people swear that they are able to set their own internal alarm clock just by banging their head on the pillow six times before going to sleep. It sounds crazy, but there is some research to suggest that it is a real thing – with the unconscious keeping track of time while we slumber.

In experiments, people who were told they were going to be woken up at 6 a.m. (and were) showed a rise in stress hormones that help us to wake up from 4.30 a.m. onwards. People who were unexpectedly woken up at the same time had no such spike in hormone activity. The implication is that the unconscious mind is not only keeping track of the hours as they tick by, but also sets a biological alarm to jump-start the waking process. Perhaps the pillow ritual helps to set that alarm.

Other research suggests that we can also understand and

process language while we sleep, if only during the earlier stages of the sleep cycle. This might explain why some words, like our names, wake us more easily than others. This makes good evolutionary sense. We are at our most vulnerable while we sleep, so rather than shut down completely, the brain goes into a kind of standby mode from which it will only wake if strictly necessary.

Subliminal mathematics

We don't have to be asleep for the unconscious to do processing on our behalf, however. Even complicated feats of reasoning, such as reading and basic mathematics, are possible without conscious awareness. In experiments to test participants' unconscious maths skills, sums are flashed up on a screen too quickly to be processed consciously. When participants were later given the sum to work on consciously they came up with the answer significantly more quickly than those who hadn't had the benefit of a subconscious 'prime'.

It seems that there is no kind of processing that the unconscious can't take care of. Every moment the brain takes in far more information than it can process. In order to make sense of it all, the brain constantly makes predictions that it tests by comparing incoming data against stored information.

Simply imagining the future is enough to set the brain in motion. Imaging studies have shown that when people expect a sound, word or image to appear, the brain generates an anticipatory signal in the relevant sensory area.

Keeping ahead of our senses

This ability to be one step ahead of the senses has an important role in helping us understand speech. Studies have shown that the brain can use one sense to inform another. When you hear a recording of speech that is so degraded it is nearly unintelligible the words sound clearer if you have previously read the same words in subtitles. The sensory parts of the brain are comparing the speech you've heard to the speech you predicted.

The brain's crystal ball even extends to predicting how other people might behave. Research shows that just seeing someone talk for 2 seconds is enough to form an opinion about their competence, confidence and honesty, and even their sexuality, political stripes and the amount of money they have. No one knows how we do this, but it seems to be an overall body signal that is both given out and picked up unconsciously and is very hard to fake.

Tracking the body

Perhaps the most underrated skill of the unconscious, however, is its ability to keep track of where our body is in space and what it is doing at any given time. This ability, called proprioception, results from a constant conversation between the body and brain and is a key building block of our sense of self. It is thought to be the result of the brain predicting the causes of the various sensory inputs it receives – from nerves and muscles inside the body, and from the senses detecting what's going on outside the body. What we become aware of is the brain's 'best guess' about where the body begins and ends.

A lack of proprioception is rare, but can happen with nerve or brain damage. People who have lost this ability find it incredibly difficult to move, and have to relearn moving their bodies consciously, making the kinds of fluid movements most people take for granted a painfully slow and deliberate process.

Take the case of Ian Waterman, who lost proprioception after nerve damage caused by a flu-like virus in 1971. After being told he would never walk again, he slowly learned to consciously control his muscles to move his body. Decades later, it was still far from easy and he only had full control over his movements if he was looking at the relevant body part and concentrating.

DON'T THINK: WHEN UNCONSCIOUS TOOLS WORK BETTER

Unconscious processing is often thought of as an initial stage of the brain's workings, but for some tasks it is actually the best tool for the job. When working unconsciously, the brain can throw the net more widely, bringing together information from all over the brain without interference from the brain's goal-directed frontal lobes. The result can be new ideas that burst through to consciousness in an 'aha' moment of insight.

Some people seem to be better wired for this kind of thinking than others. But while there is no known way to change your brain to make it more creative, one way to access the right state is to work on a problem until you get stuck, then take a break and let your mind wander. With luck, something useful will bubble up before your deadline.

THE WEIRD WORLD OF YOUR BRAIN ON AUTOPILOT

IN 1953, a physician named Louis Sokoloff laid a twenty-year-old college student on a hospital trolley, attached electrodes to his scalp and inserted a syringe into his jugular vein. For 60 minutes the volunteer lay there and solved arithmetic problems, while Sokoloff monitored his brainwaves and checked the levels of oxygen and carbon dioxide in his blood. Sokoloff expected to see his volunteer's brain guzzling more oxygen as it crunched the problems, but was surprised to find that it consumed no more oxygen while doing arithmetic than while he was resting with his eyes closed. So what was the brain doing?

It wasn't until the 1990s that brain scans revealed that far from taking a break, the brain keeps on chugging away when we rest our conscious minds. But for a long time exactly what it was doing while our minds were elsewhere has remained a mystery. We now know it has a kind of 'autopilot' mode, which enables us to carry on background tasks quickly, efficiently and without conscious thought. Unlike a computer on standby waiting to be given a task, the brain uses its downtime to get on with a spot of housekeeping.

Brain-imaging studies have revealed that the brain's autopilot mode is the job of a set of brain structures arching through the middle of the brain, from front to back, and collectively known as the default mode network. This group of brain regions seems central to many of the key functions of the brain – so much so that some parts of the default mode network devour 30 per cent more calories, gram for gram, than most areas of the brain.

So, what is it doing with all that energy? Some clues to its purpose come from the specific brain areas that are linked up in the network. One of the core components is an area called the medial prefrontal cortex (MPC), which is known to evaluate whether a situation is likely to be good, bad or indifferent from the person's point of view. People who suffer damage to their MPC become listless and uncommunicative. One woman who recovered from a stroke in that area recalled inhabiting an empty mind, devoid of the wandering, stream-of-consciousness thoughts that most of us take for granted. Parts of the default mode network also have strong connections to the hippocampus, which records and recalls autobiographical memories such as yesterday's breakfast or your first day at school.

This all points to one thing: daydreaming. Through the hippocampus, the default mode network could tap into memories – the raw material of daydreams. The MPC could then evaluate those memories from an introspective viewpoint, providing the brain with an 'inner rehearsal' for considering future actions and choices. Daydreaming may sound like a mental luxury, but its purpose is deadly serious: it is the ultimate tool for incorporating lessons learned in the past into our plans for the future. So important is this exercise, it seems, that the brain engages in it whenever possible, breaking off only when it has to divert its limited supply of blood, oxygen and glucose to a more urgent task.

The default mode network may also be involved in selectively storing and updating memories based on their importance from a personal perspective – whether they're good, threatening, emotionally painful, and so on. To prevent a backlog of unstored memories building up, the network returns to its duties whenever it has a spare moment. Hence the constant chatter with the hippocampus. It also seems to play an important role in learning, allowing us to move from deliberate thought and action to working on autopilot. When people learn the rules of a new game in a brain scanner, their brain activity resembles patterns that are typical of learning minds. But as they become more expert, the default mode network becomes more active – and responses faster and more accurate. This suggests that when we 'switch off', our brains go into an autopilot mode that allows us to perform tasks without thinking much about them.

This might also help explain why some tasks – such as playing a well-known tune on a musical instrument or dialling a phone number that you call often – suddenly seem much more difficult when you go from doing them absent-mindedly to consciously thinking about them.

Network disruption

The default network's pattern of activity is disrupted in patients with Alzheimer's disease and in other maladies including depression, attention-deficit hyperactivity disorder (ADHD), autism and schizophrenia. It also plays a mysterious role in victims of brain injury or stroke who hover in the grey netherworld between consciousness and brain death known as a minimally conscious or vegetative state. Understanding it better may help in the hunt for better treatments for these conditions.

AUTOPILOT UNDER CONTROL

When Zen Buddhists meditate, they may be deliberately switching off their default network, the system within the brain that has been strongly linked with daydreaming. The goal of Zen meditation is to clear the mind of stream-of-consciousness thoughts by focusing attention on posture and breathing. A group of volunteers trained in Zen meditation were put into an fMRI scanner, presented with random strings of letters, and asked to determine whether each was an English-language word or just gibberish.

Each time a subject saw a real word, their default network would light up for a few seconds – evidence of meandering thoughts triggered by the word, such as apple . . . apple pie . . . cinnamon. Zen meditators performed just as well as non-meditators on word recognition, but they were much quicker to rein in their daydreaming engines afterwards, doing so within about 10 seconds, versus 15 for non-meditators.

ACCESSING THOUGHTS YOU DON'T KNOW YOU'RE HAVING

IT'S ALL VERY well having a clever unconscious running the show, though it would be nice to know what it's up to now and again. But since, by definition, we are not aware of what is going on outside our consciousness, accessing our unconscious musings is always going to be hard.

There are, however, a few tried and tested ways to eavesdrop on your hidden thoughts. Harvard University's Project Implicit shines a light on people's unconscious biases using quick-fire questions that assess how readily they associate words such as 'black' and 'white' with others such as 'good' and 'bad'. The project's website has a slew of online tests exposing unconscious attitudes to race, gender and homosexuality. The questions flash up fast so it's hard to cheat. Try them – you may be surprised at what your unconscious has to say. Once you have a better idea of what is going on in the depths of your automatic thoughts, it is much easier to step in and challenge them.

Mind reading

Psychologists also use a method called experience sampling. Pioneered in the 1970s by Russell Hurlburt, a psychologist at the University of Nevada, Las Vegas, the method prompts its volunteers to record their inner experiences at random intervals throughout the day. If you think of consciousness as a spotlight on the dark room of the unconscious, experience sampling directs the light at what is usually hidden away in the corners and brings it into the light for inspection.

Volunteers are asked to wear an earpiece linked to a beeper, which goes off at random intervals six times a day, prompting them to note their thoughts. At the end of the day there is an hour-long interview to tease out what people are thinking and how. This method has revealed that much of the time our unconscious mental chatterings are fairly inconsequential. Even when done with famous scientists, there has been very little evidence of genius or world-changing ideas. What it has shown, though, is that we all experience the world slightly differently. Some people hear their thoughts in words, while others see a visual image in their mind's eye. Others might feel emotions physically or experience the world via colours or sensations. And some get a mix of all of several experiences at once. Many of our unconscious experiences may even underlie common turns of phrase. When some people say they see red when they get angry, for example, they may literally see the colour red in front of their eyes. And when someone is stressed and says their head is spinning, they may physically feel dizzy.

Strangely, most people have no idea about how they think until they take part in this kind of experiment. Almost everyone thinks that they think in words, for example, but it is rarely the case. After four decades of doing this, Hurlburt has come to the conclusion that most people have no idea what is running through their minds, but that they can be taught to tune in to it in just a few days.

There are a few smartphone apps, such as iPromptU, that

will enable you to record your thoughts in this way, but another option is learning to meditate. Recent studies suggest that expert meditators have better access to their unconscious than most people. In a classic experiment in free will (see What if you have no free will?, page 239) people typically feel like they decide to press a button about 200 milliseconds before their finger moves. Electrodes attached to their skull, however, reveal that activity in the part of their brain that controls movement occurs a further 350 milliseconds prior to their decision. Probably this means that in fact it is the unconscious brain that 'decides' when to press the button. When the experiment was done with regular meditators, though, the meditators had a longer gap in time between when they felt like they decided to move their finger and when it physically moved. This suggests that, not only were they recognising the unconscious brain activity that precedes movements and decisions earlier than most people, but that the ability to do so can be trained.

Non-meditators were also tested on how well they could be hypnotised. After they came out of any hypnotic trance the experiment was repeated. Those who could be easily hypnotised felt like they decided to move their finger 124 milliseconds later than did those of low hypnotisability. In fact, the easily hypnotisable group had the sensation of deciding to move 23 milliseconds after their finger had actually moved.

It is not that people who are highly hypnotisable are puppets, but that they may have less conscious access to their unconscious intentions. Previous research has suggested that people who meditate are less easy to hypnotise and people who can be hypnotised are less 'mindful'; in other words they are less aware of their internal bodily processes.

Another study using the same set-up has shown that people

who are impulsive also have shorter time intervals between their conscious awareness of an intention to act and the act itself. So getting some control of your unconscious might be as simple as learning to be a little more present.

TRUST YOUR GUT

We take in far more information unconsciously than we ever notice. But that doesn't mean that the information isn't in there at all. In experiments where people report that they have no conscious recollection of having seen a particular face, they are able pick it out of a group of faces far more often than would be expected by chance. Even though they didn't remember seeing the person, the information was in their short-term memory.

This could be useful to know. Imagine you are in a café wanting to pay your bill and you look for the waiter who initially took your drinks order. It may be hard to recognise them, but if you pay attention to your gut instinct, you will probably get it right. Your subconscious may have registered the information on your behalf.

HOW TO HACK YOUR BRAIN'S AUTOPILOT

OLD HABITS DIE hard. But they are also incredibly useful. As much as 40 per cent of our daily behaviour, from brushing our teeth to commuting to work, requires no mental input at all. It's just as well: giving everything our full attention would become exhausting.

Habits, though, have a downside. Many of our least healthy behaviours just happen, without us making any conscious choice to engage in them. And as anyone who has ever tried to stop biting their nails or quit smoking will know, once they have become entrenched, habits are fiendishly difficult to break.

Scientifically, habits are defined as actions performed routinely in certain contexts, often unconsciously. Once a habit is formed, it runs on autopilot, saving both time and mental effort. Advances in neuroscience have made it possible to peer inside the brain as it goes about its business of making and breaking habits, and to find out what happens as a new habit is formed. This opens up the possibility of finding ways to switch good habits on and bad ones off.

Habits in brackets

A key player in the development of a habit is the striatum, a region important for movement, mood and reward. When a rat learns to navigate a maze and begins to follow the same route out of habit, brainwaves slow down in this part of the brain, suggesting that brain activity in that region has become more coordinated and efficient. Something similar happens in monkeys, and perhaps humans too. Importantly, studies showed that cells within the striatum fire in this way at the beginning and end of a behaviour, as if signalling when the autopilot program is turned on and off. It's almost as if the brain is putting the habitual action between brackets.

This makes the process more efficient, but the downside is that once the brackets have opened, the unconscious runs the rest of the program with no further input. That's why simply wanting to stop biting your nails isn't enough. You need to find a way to close the brackets early.

We know this can be done because in experiments where a small brain area called the infralimbic cortex is removed in an animal it led them to abandon their habits and act in more goal-directed ways. Using optogenetics, a precise technique that allows neurons in this region to be turned on and off with flashes of light, it was possible – in rats – to turn habitual behaviours on and off at the flick of a switch.

This raises the intriguing possibility that targeting certain brain areas could help us break bad habits. Optogenetics hasn't been tried on human brains yet, but transcranial magnetic stimulation – a non-invasive technique that applies small electromagnetic currents to the outside of the head – is a possible alternative. Deep brain stimulation – in which an implanted electrode is used to activate a certain brain region – is another, more drastic option. It is already used to treat severe cases of depression and Parkinson's disease, although studies on obsessive–compulsive disorder (OCD), which is associated with very persistent, habit-like behaviours, have so far been mixed.

While we wait for an easy way to rewire our brains for better habits, willpower is the best tool we have. Unfortunately, the brain's goal-directed system, which keeps us on the straight and narrow, takes a lot of effort and mental resources to run. When it gets tied up, as during a week of exhausting exams, the effortless habit system kicks in. This is a double-edged sword, as it increases all kinds of habits, good and bad. You might eat badly in times of stress but you might exercise more too.

If you want to maximise the good habits, it's a good idea to get to know the things that set them off. We know that habits are triggered by certain cues or contexts, which offers the possibility of hacking them to your advantage. So, if you want to drink more water, put a jug on your desk. If you want to walk instead of drive, hide your car keys.

Clean break

The link between habits and our environment is also why the best times to break habits or create new ones are when we go on a trip, change jobs or move house. Once the context gets disrupted, old habits fade and new ones can form.

And don't worry about little slip-ups. If you diet for a month, but then fall off for a day, don't take that to mean you've failed. To prevent slip-ups from snowballing, break the day into sections. That way, if you eat too many doughnuts at a morning meeting, you can start afresh at noon and try again.

The good news is that it gets easier. Willpower is like a

muscle; although it can get depleted, it also gets stronger with practice. And, if you can master your brain's autopilot system, you can make a habit of whatever you want.

DON'T QUIT: YOU MAY HAVE MORE WILLPOWER THAN YOU THINK

The idea that willpower is a limited resource has been received wisdom in psychology circles for nearly twenty years. According to a series of recent findings, though, our levels of self-control may not be a budget we have to eke out, but a renewable resource that can be powered up as we go along.

According to the idea that willpower is limited, the difference in people's ability to stay strong in the face of temptation can be explained by the amount of fuel in our mental reserves. Recent studies, however, suggest that this only happens if you already believe that willpower is limited. In experiments, people who believe that there are no limits to our willpower showed less signs of mental fatigue after a taxing task than those who saw mental energy as a finite resource. Offering cash for greater effort can also prevent a loss of willpower, suggesting that our wellsprings of willpower are deeper than we think. The message emerging is that where there's a will there's a way. You just have to believe that you can do it.

06.

THINKING

HOW WE THINK

'I THINK, THEREFORE I AM', wrote philosopher René Descartes in 1637. Our ability to think has long been considered central to what makes us human. Thinking is at the heart of everything we do. Thoughts – profound, mundane, logical or bizarre – pervade our every moment. But what does 'I think' really mean? Getting to the bottom of our thought processes is a tough challenge.

A good starting point is to think about the difference between thought and perception. Suppose there is a sandwich in front of you. You can see it and smell it, but that's purely a perceptual event. But if you think about the sandwich you ate last week, or how the bread was made – these are thoughts. Thought can extend our mental reach into times, places and ideas that are inaccessible to our senses. We can think about things we can't see (such as a black hole), and even things that don't physically exist (such as the number pi), things that happened long ago, and events that are imagined or in the future.

Unconscious ruminations

Things are complicated further by the fact that thoughts are often unconscious, springing to mind when you're thinking of something else. You don't need to think of something consciously for your brain to create thoughts about it. Thoughts are private too, and we're not even aware of much of the stuff that goes on in our heads. So how can these ruminations be studied or understood?

One way is to ask people to write down their thoughts at random intervals in the day, in response to a beep. Psychologist Russell Hurlburt has been doing just this since the 1970s. It is actually surprisingly hard to tune in to your internal chatter, but most people can get the hang of it within a few days. As you might expect, each person's thoughts are highly individual, and vary according to what they are doing, but this is one example:

> Emma is cleaning up her kitchen . . . she had picked up a glass from her counter and noticed two vases of mostly dead flowers . . . she's putting the glass upside down into the dishwasher; the seeing of the glass and surrounding glasses and trying to get it to fit are part of her experience . . . internally she hears the crunchiness of dead flowers (as if she has picked up some fallen rose petals) . . . Simultaneously internally she sees the two vases of flowers and the countertop with the dead petals. This imaginary seeing is in colour and detail . . . She has a sense that the cleaning up of the flowers can wait; the sense of 'can wait' is somehow present without words or symbols.

Categorising these thoughts is practically impossible. In this example, is Emma thinking about clearing up the kitchen? Dead flowers? Or nothing in particular? Nevertheless, Hurlburt has identified five basic modes of thinking. The first is inner speaking, where you hear words in your own voice. Then there's inner seeing, in which a visual image appears in your mind's eye. Our thoughts can also take the form of sensory awareness, or a feeling which arises from any strong emotion. This can even be imagined, thinking of something soft, for example. Finally, there is unsymbolised thinking, in which a concept is unattached to any mental words or images.

One thing that stands out is how humdrum these thoughts are, in stark contrast to the deep and profound nature that most people associate with the act of thinking. The classic portrayal of thinking is Auguste Rodin's famous statue *The Thinker*, depicting a man with his chin resting on his hand. He appears to be lost in deep, deliberate thought, but the kind of thinking that Rodin depicted is actually a very narrow class of thought: systematic, logical and goal-directed. This is a useful and powerful tool – and we rightly praise people who are good at it, admiringly calling them 'thinkers'. But to define thought this narrowly ignores the true richness of our mental lives. A significant amount of our thought is not directed at a specific goal and it is common for our minds to wander away from the task at hand.

Tumbleweed in the mind

Think about thinking – in itself a mind-boggling feat – and you quickly realise that it is mostly a passive and undirected activity, like a tumbleweed blowing around an empty street (you probably now have an image of that in your head, assuming you are not one of those people who doesn't have a mind's eye). And that thoughts come in many varieties, from idle reverie to determined problem-solving. As far as we know, these are experiences only humans enjoy.

But try to articulate where thought comes from, and you may find yourself stymied. Thinking comes so naturally to us that we rarely stop to consider how ineffable and ill-understood it is. Nor is it any easier to pin down how we solve problems, or come up with ideas.

We have developed many techniques that help us do these things, from logic to brainstorming to philosophy. And there are technologies to help gather individual thoughts into collective knowledge, from writing to social networking. The combination of thinking techniques and information technologies has helped us make great strides. It is hard to envisage how our technological and intellectual achievements would exist without it.

OBSESSIVE THOUGHTS

Sometimes our minds constantly meander back to familiar territory. Many people obsessively count in their head, for example, as they climb steps or put things away. Why?

One explanation is that this is a form of 'mental doodling', a way for the brain to keep active during boring or repetitive tasks. An alternative explanation is that these default thoughts represent a kind of mental ritual. There are times when we all experience this kind of obsessional thinking – normally between the ages of two and four. Toddlers often develop ritualistic behaviours like not walking on cracks in the pavement, and need rituals related to bed- or mealtimes, perhaps as a way to help them make sense of a world in which much is new. At that age, we need order, we like to have things the same – a desire that some of us never lose. Obsessional thoughts only become a problem when they are intrusive, causing distress or functional impairment.

YOUR HARE AND TORTOISE
THINKING SYSTEMS

SOMETIMES THE ANSWER to a problem springs to mind from nowhere. Without any conscious effort, your brain just comes up with the solution. At other times it's not so easy. Only after laboriously working through the permutations do you finally arrive at the answer. What's going on?

The brain has two different systems that control the way we think and make decisions: one fast and one slow. These correspond roughly to a distinction everybody is familiar with: thoughts that occur to you and thoughts that you have to generate. The Nobel-prizewinning psychologist Daniel Kahneman and his colleague, the late Amos Tversky, conducted decades of research into these ways of thinking, and dubbed them 'system one' and 'system two'.

When you recognise an object or have a sudden feeling that you like someone, these belong to system one thinking – fast, automatic, mostly effortless. System one is the mental ninja with lightning reflexes, using a limited amount of information to get to conclusions in a fast and shallow way – worth it in dangerous situations when slowness equals death. System two is the professor. It thinks deeply and theoretically, taking everything into account and producing much mental pain –

worth it when avoiding error is what matters most. It is slow, deliberate, effortful thinking.

Both have drawbacks. System one might come up with quick answers, but it relies on habits that are slow to adapt to new circumstances. It can be triggered by associations in our memory, and we have little control over this. Our perception of the world isn't always accurate and many of our errors of judgement are due to limitations of system one; whereas in system two the limitations are mainly down to the breadth or otherwise of our knowledge. System two is also slow. You couldn't possibly live your life deliberating all the time.

You can avoid certain errors of thinking by being aware of these limitations. For example, suppose you are asked the following question: is the average price of German cars more or less than £90,000? Afterwards, you are asked to estimate the average price of German cars. Your answer will be quite different from what it would have been if you had first been asked: is the average price of German cars more or less than £12,000. It will be much higher. This is because when you are given a ridiculously high number, you automatically think of expensive cars, such as Mercedes or BMWs. If you are given a low number, you'll probably think of cheaper brands of cars. Your mind automatically generates a biased sample.

This phenomenon is called anchoring. It causes you to focus on a given number, and that number becomes plausible just because it's been mentioned. A lot of what happens in negotiations are attempts of one side to anchor the thinking of the other. Another thing to watch out for is jumping to conclusions. System one produces the best coherent story possible from the evidence at hand. In fact, it is a machine for jumping to conclusions. But this trait can also lead us astray.

If you are asked whether X is a good leader and you are told that she is intelligent and strong, you have already formed the impression she's a good leader. But you haven't been told other things – she's also corrupt and cruel, say. You haven't waited for more information but formed an impression on the basis of the information you had. We can't live in a state of perpetual doubt so our mind makes up the best story possible and we live as if this story were true. Often we are not aware of how little information we have, and if we don't realise this then we get the phenomenon of overconfidence. Confidence is not a judgement, it's a feeling.

Mistaken hunches

Another failing of our brain is its ineptitude at understanding probabilities. Suppose, for example, that you have an ordinary six-sided die with four green faces and two red ones. Your task is to choose one of the sequences below, then roll the die twenty times. If your selection turns up, you will win £25. Which one will maximise your chances of winning?

1. RGRRR
2. GRGRRR
3. GRRRR

A majority of people choose (2) over (1). But that cannot be right: sequence (2) contains (1), so (2) must be less likely than (1). These, and many other mistakes are endemic in human thought. Why? When you look at the three sequences, you see that the mix of green and red in (2) is closer to the proportions of green and red on the die itself. A 'representativeness heuristic' programmed into your mind tells you that representative choices are good choices; you therefore select (2).

You are not the helpless victim of the representativeness heuristic, however. You can also think your way carefully through the problem and see that although sequence (2) is more representative than (1), it is less probable. Why, then, do so many people make the wrong choice? Because thinking using the representativeness heuristic is fast, effortless and automatic, whereas thinking logically is slow, relatively difficult and requires determination and self-control. Everyone feels the pull of choice (2), but only those who push themselves to double-check will work out the correct answer.

But beware of overconfidence in overcoming these slip-ups. While we can try to force our brains to chew over a problem slowly, our gut feelings are hard to ignore. Even Kahneman, who has studied the errors of human thinking for decades, says his intuitions have not changed.

BEYOND FAST AND SLOW

Fast and slow thinking are just the tip of a cognitive iceberg. Recognising what lies beneath means learning to live with unsettling and counter-intuitive ideas about how our brains work. Take a statement as simple as 'Ann approached the bank'. Has she got a loan on her mind? Was she walking down the high street?

It's only when you know that the preceding sentence had been 'They were floating gently down the river' that all

---

becomes clear. Making sense of a situation depends on context, and is further complicated because it depends not only on visual cues, memories and associations, but also on our goals and anxieties.

TEST YOUR LATERAL THINKING

BEING SMART IS not always about the kind of intelligence measured by IQ tests. Often a strong dose of lateral thinking is required – your brain has to shift gear and look at things from a different angle. You can test this out with the following problems.

1. What five-letter English word contains four pronouns?
2. This group of words is most unusual. Why? If you look at it, you will possibly find out, but a solution is not all that obvious.
3. The letters **O, T, T, F, F** . . . form the beginning of an infinite sequence with a simple rule. What are the next two letters of the sequence?
4. Two English words begin with the letters **HE** and end with the letters **HE**. Can you find them both?
5. What positive fraction smaller than one is equal to itself when inverted?
6. Using six line segments of equal length, can you construct four equilateral triangles, such that the sides of the four equilateral triangles are the same length as the line segments?

7. Consider the letters **H, I, J, K, L, M, N, O**. The solution to the problem is one word. What is the word?

8. The numbers **8, 5, 4, 9, 1, 7, 6, 3, 2** form a sequence. Find the rule for the sequence.

9. Two fathers gave their two sons some money. One gave his son 150 rubles and the other 100 rubles. When the two sons counted their finances, they found that together they had become richer by only 150 rubles. What is the explanation?

10. What three-letter acronym contains four letters?

11. What is the missing letter? **H, Z, X, O, I, S.**

12. The meaning of a common English word becomes plural when an A is added at its start. What is the word?

13. Can you think of three plural nouns that become singular when 's' is added?

14. In her work, my friend doubles the number of people that she sees every day. What does she do?

15. What five-letter word becomes shorter when you add two letters to it?

16. What is remarkable about this sentence?
    Show this bold Prussian that praises slaughter, slaughter brings rout.

17. And what about the following one?
    I am not very happy acting pleased whenever prominent scientists overmagnify intellectual enlightenment.

See page 411 for the answers.

HOW LANGUAGE SHAPES YOUR THOUGHTS

AS YOU READ this article and your eyes follow the words across the page, what's going on in your head? You may be aware of a voice in your head silently muttering along. You are not alone in your internal babbling. Measuring the contents of people's minds is difficult, but it seems that up to 80 per cent of our mental experiences are verbal.

What do these words do? Are they the basis of our thoughts? Or is language simply the tool we use to communicate our thoughts? This is the subject of intense debate, but a picture is beginning to emerge. Language helps us to think and perceive the world. Words bring a smorgasbord of benefits to human cognition, from abstract thinking to sensory perception.

The idea that language guides human thinking and shapes perception has a long history. Philosophers have toyed with it for centuries, but modern psychologists are putting flesh on its bones. For a long time, our 'inner speech' was believed to be outside the realms of science, but now researchers are getting a handle on how it is created in the brain. If it is derived from external speech, for example, then both external and inner speech might be expected to activate the same neural networks. Brain-scanning studies have confirmed this. There are also hints that inner speech is just external speech without articulation. Neuroscientists are zoning in on that process, picking up the neural signatures of imagined speech – that inner voice – using brain scans.

So much for the subjective qualities of inner speech. What, if anything, does it actually do? One idea is that inner speech acts as a tool to transform a mental task, just as the use of a screwdriver transforms the task of assembling a shed. Putting our thoughts into words gives them a more tangible form, which makes them easier to use. It may also be that verbal thought can allow communication between other cognitive systems, providing a common language for the brain.

When the word doesn't exist

One contentious area is the role that language plays in shaping our thoughts. Key evidence comes from studies of the Pirahã, a tribe of hunter-gatherers in Brazil. Their language doesn't contain words for precise numbers, and tests have shown that they are unable to tell reliably the difference between four and five objects placed in a row. For some, this provided some of the strongest evidence that the language available to humans defines our thoughts. When a language lacks words for certain concepts, it could actually prevent speakers of the language from understanding those concepts.

Another study suggested that Zuni Native Americans, who use the same word for yellow and orange, have more difficulty remembering whether an object is yellow or orange than English speakers. And Russian speakers, who have two words for different shades of blue, really are faster at discriminating between the different shades than English speakers.

But the evidence is not always clear cut. The Dani people of

New Guinea, for example, who have just two colour terms: light and dark, could tell the difference between the hues of different objects just as effectively as English speakers.

Some languages do seem to influence how their speakers think about things like space, time and even emotions. Certain cultures will name an emotion, for example, and once it is named, it is noticed more often. People then feel it more, because they have a word for it. A good example is the Japanese word *amae* – something that is like the comfort of knowing you can depend on another's support. The word might have emerged from an expressive need in Japan's collectivist culture, but once in existence it might also have allowed that aspect of culture to flourish.

Infants can more easily group objects into categories – animals versus vehicles, say – if they have already learned the category names. Other research suggests that the spatial reasoning of young children is improved by reminding them of words such as 'top', 'middle' and 'bottom'. Meanwhile, a few studies have described how people who lost their language skills following a stroke have struggled with tasks such as grouping and categorising objects.

Experience shaped by words

Perhaps the most surprising effect of language is the way it shapes perception. The words you say, think and hear do seem to have a very real impact on the way you see. Hearing verbs associated with vertical movement – such as 'climb', 'rise' or 'drip' – affects the eye's sensitivity to such motion. In one study, volunteers were shown a display consisting of a thousand dots, each of which moved either vertically or randomly. They were more likely to detect the predominant direction of motion when they heard a verb that matched it, for example 'rise' when most of the dots were moving upwards. Conversely, they were less likely to detect the movement if the verb described the opposite motion, such as 'fall' when the dots were rising.

While researchers are still gathering the evidence, these results certainly suggest that the voice in your head is important to many cognitive processes. When language emerged, deep in our evolutionary past, it seems that our inner voice changed the way we experience the world.

NO INNER VOICE

If language is so important for thinking, then what about people who, for various reasons, don't talk to themselves in the usual way? As you might expect, deaf people who communicate in sign language often talk to themselves in sign language too. People with autism, meanwhile, who often have problems with linguistic communication, seem not to use inner speech for planning, although they do use it for other purposes such as short-term memory. A more dramatic difficulty comes from damage to the language areas of the brain, which can silence some people's inner voices. One such individual, neuroanatomist Jill Bolte Taylor, reported a lack of self-awareness after a stroke had damaged her language system – supporting the view that verbal thinking may be important for self-understanding.

WHAT IF YOU CAN'T THINK IN PICTURES?

PICTURE, FOR A MOMENT, a sunny beach: shimmering blue water, waves rolling on to the shore, colourful umbrellas dotted along the sand. For some people, this act is impossible; they are unable to 'see' anything in their minds. When it comes to mental imagery, they are blind.

Take Craig Venter, the biologist who led one of the teams that first sequenced the human genome. He attributes his academic success to an unusual way of thinking, using purely concepts with no mental imagery whatsoever. He says it's like having a computer store the information, but without a screen attached to the computer.

For most of us, mental imagery plays a significant role in our thought processes. So do people whose mind's eye is blank think differently? Science is starting to find answers. And studying people with this condition is helping to reveal a lot about how our brains process the things we see around us.

We have known of the existence of people with no mind's eye for more than a century. In 1880, Francis Galton conducted an experiment in which people had to imagine themselves sitting at their breakfast table, and to rate the illumination, definition and colouring of the table and the objects on it. Some found it easy to imagine the table, including Galton's cousin, Charles Darwin, for whom the scene was 'as distinct as if I had photos before me'. But a few individuals drew a total blank.

Mental imagery

Today there is a standard way to probe the acuity of the mind's eye: the Vividness of Visual Imagery Questionnaire. It asks people to imagine various scenes and rate the clarity of the mental picture. Surveys show that most people have fairly vivid mental imagery; only 2 to 3 per cent report a completely image-free mind. For a long time, no one gave much thought to what caused this. That changed in 2003, when neurologist Adam Zeman studied a patient known as MX who reported losing his mind's eye after heart surgery. Zeman decided to find out what was going on inside MX's head.

We have a good idea how creating a mental image usually works. When you see a real object, the information captured by your eyes and fed to the brain activates a pattern of neurons unique to that object: a chair has one distinct pattern, a table another. MRI brain scans show that when you imagine a picture of that object, the same neural pattern lights up, just slightly less strongly than when you are actually seeing it.

To find out how MX's brain worked, Zeman put him into an MRI scanner and showed him pictures of people he was likely to recognise, including Tony Blair, the former UK prime minister. The visual areas towards the back of his brain lit up in distinctive patterns as expected. However, when MX was asked to picture Blair's face in his mind's eye, those areas were silent. In other words, the visual circuits worked when they had a signal from the outside world but MX couldn't switch them on at will.

But then came an unexpected finding. Even though MX couldn't form a picture of Tony Blair, he could handle tasks that would seem to require one – stating Blair's eye colour without seeing a picture of him, for example. He also aced other tests, such as imagining standing in his own home and counting its windows (even though he had no mental image of his house, he had an awareness of being there).

Soon after Zeman published his results, he heard from another twenty-one people who said they had this condition, which he called aphantasia. However, unlike MX, they claimed to have had it from birth. A battery of cognitive tests soon confirmed they had the condition and that, like MX, they had no problem getting on with life, including tasks that might seem impossible without a mind's eye.

Seeing without seeing

It might sound paradoxical, but these 'tests of visual imagery' aren't always difficult to complete without a mind's eye. Take the window-counting test. People with aphantasia might not experience a mental image of their house, but rather an awareness of being there. Craig Venter says it is the same for him. He doesn't have to 'see' events to relive them, he says. There are different ways of storing visual information other than with a picture.

How? Visual imagery is not constructed in just one way in the brain. There are separate circuits for things like shape, colour and spatial relationships, among much else. Take the following question: in the upper-case letter 'A', what shape is formed by the enclosed region? A person with aphantasia won't be able to picture this letter in their mind, but might arrive at the right answer by imagining that they are drawing the letter.

This gives us a clue as to how aphantasics deal with apparently pictorial information. To complete the task they piggyback on neurons involved in controlling physical movements rather than using the visual brain circuitry. They might be able to imagine a letter not because they can 'see' it, but because they can imagine making it.

Mental pictures aren't the only way to process 'visual' information, they might not even be the best way. Zeman has been contacted by a number of aphantasic artists. You would assume that artists in particular need a mind's eye, but perhaps this isn't the case. Then there is Venter, who sees a connection between his aphantasia and his scientific achievements. Perhaps not having a mind's eye forces you to see the world differently, resulting in an unusual eye for art or alternative modes of thinking.

MIND READING

One way to unpick our thoughts is to focus on the electrical signals that create them. Our brains work in unique ways, and the way each of us thinks about a concept is influenced by our experiences and memories. This results in different patterns of brain activity that neuroscientists can look at and understand. It's early days, but mind-reading devices are already in development that allow locked-in people, who are completely paralysed except for their eye movements, to communicate or control wheelchairs using thoughts alone.

WHY IT'S GOOD THAT
I KNOW THAT I KNOW

MURPH, WHO IS TEN, sits in front of his computer screen playing a game that consists of classifying images. He finds out how he is doing by the sound of happy whoops for getting it right and buzzers for getting it wrong. He doesn't seem to like being told he is wrong but has learned that he can avoid this by choosing to pass when he is not sure of the correct answer. Natua plays a similar game, but he has to discriminate between sounds, choosing whether they are high or low-pitched. He doesn't like being wrong either, and when the task gets really tough he also chooses to pass rather than guess.

We all know that feeling of not knowing. When an answer pops into our heads we get a buzz of confident recognition, an 'aha' moment. If we don't know, we might feel a frustrating tip-of-the-tongue sensation, a mild form of panic or embarrassment, perhaps even gut-wrenching anxiety when we realise that we don't have a clue. Maybe that explains why Murph and Natua would rather pass than admit they don't know. People do this all the time, so what's the big deal? Well, Murph is a monkey, and Natua a dolphin. Their game-playing has got some researchers very excited because it suggests that these animals might know something about what's on their own minds.

Knowing what you know, and what you do not, may not sound that clever, but it is a very important mental skill. Philosophers have long debated the significance of being able to think about thinking or know about knowing. This type of abstract thought, which is called metacognition, could be an important first step on the ladder of consciousness: knowing what's on your mind, to reflect on something that is mental, not in the environment, might be a prerequisite for knowing about whose mind it is, which is important for having a concept of self, self-awareness and, eventually, full reflective consciousness.

Thinking about what you know

In practical terms, the ability to think about what you know is a big advantage. Getting things wrong can be costly, even if it is just wasteful of time, and metacognition allows you to pause, reflect and seek more information if you need it. With something so useful, you might assume other animals would have the ability too. This is a hotly debated topic. It is still often assumed that abstract thought needs some form of language. No wonder the apparent discovery of metacognition in monkeys and dolphins rattled a few cages. But more recently, rats, bees and even ants have also been found to show signs of metacognition.

Humans, however, take metacognition to much higher levels. One of the most obvious differences between human thinking and that of animals is our level of self-awareness. A pet dog, for instance, is probably aware of many sensations at any given moment: that it is hungry, that it is tired after a long walk, perhaps, and that there is a delicious smell emanating from the kitchen. Its owner would be aware of sensations like those and yet have an extra level of thought processes overlaying them. As a human, we can be aware that we are aware of our basic sensory inputs, and that allows us to reflect on the accuracy or validity of our feelings and judgements. That lets us think: 'How tired I am after that long walk, it's that satisfying kind of tiredness you get after exercise. But I'm not too tired to walk to the pub tonight.'

How does our brain carry out these complex thought processes? Scanning the brains of humans while they carry out metacognitive tasks suggests the seat of this ability lies in our prefrontal cortex, at the front of our heads. But this faculty has been hard to measure. If we ask people how sure they are about their answers in a test, say, the results are muddled by the variation in people's ability to do the test. So are you measuring ability or awareness of that ability?

A clever test can tease this apart. It's a simple visual task, showing people stripy patches in different shades of grey, and asking which had the greatest contrast. After each question, subjects had to rate how confident they were that they had chosen the right answer. Crucially, the contrast of the stripes was adjusted for each person so that, no matter how good their vision, everyone got about 70 per cent of the answers right. This meant that for the confidence ratings, the only variable was people's metacognitive abilities, which were found to vary widely.

As well as doing these tests, the volunteers also had their brains scanned, and this revealed that those with the best metacognitive abilities had more grey matter in an area at the very front of the prefrontal cortex, known as the anterior prefrontal cortex. This lies just behind the forehead. What is it about this region that gives us this ability? It is more developed in humans, and this could mean that we have a fundamentally different self-awareness from that of animals.

It is possible that impairments of metacognition may be involved in disorders such as schizophrenia, which involves delusions and hallucinations. Schizophrenics have a problem with that very central metacognition, that 'I know I'm me and I know what I'm doing'. A better understanding of metacognition could one day help people with this condition. It may also be possible for us all to improve our metacognition through training, to accurately reflect on what you see, or whether you just made a good decision.

EXTREME FAILURE OF METACOGNITION

One of the starkest failures in metacognition is caused by a mysterious rare condition called blindsight, usually associated with brain injury. Those affected act as though they are, to all intents and purposes, sightless. But careful testing reveals they can take in some visual information about the world at an unconscious level. When asked to guess what object is in front of them, for instance, they do better than if they had just guessed randomly – insisting all the while that they can see nothing.

YOUR BRAIN IS NOT THE ONLY ONE DOING THE THINKING

IT'S BEEN A tough morning. You were late for work, missed a crucial meeting and now your boss is annoyed with you. At lunchtime you walk straight past the salad bar and head for the stodge. You can't help yourself – at times of stress the brain encourages us to seek out comfort foods. That much is well known. What you probably don't know, though, is that the real culprit may not be the brain in your skull but your other brain.

Yes, that's right, your other brain. Your body contains a separate nervous system that is so complex it has been dubbed the second brain. It comprises an estimated 500 million neurons – about five times as many as in the brain of a rat – stretching from your oesophagus to your anus. It is this brain that could be responsible for your craving under stress for chocolate and biscuits.

Embedded in the wall of the gut, the enteric nervous system (ENS) has long been known to control digestion. Now it seems it also plays an important role in our physical and mental well-being. It can work both independently of and in conjunction with the brain in your head and, although you are not conscious of your gut 'thinking', the ENS helps you sense environmental threats, and then influences your response.

If you look inside the human body, you can't fail to notice the brain and its offshoots of nerve cells running along the spinal cord. The ENS, a widely distributed network of neurons spread throughout two layers of gut tissue, is far less obvious, which is why it wasn't discovered until the mid-nineteenth century. Digestion is a complicated business, so it makes sense to have a dedicated network of nerves to oversee it. As well as controlling the mechanical mixing of food in the stomach and coordinating muscle contractions to move it through the gut, the ENS also maintains the biochemical environment within different sections of the gut, keeping them at the correct pH and chemical composition.

But there is another reason the ENS needs so many neurons: eating is fraught with danger. Like the skin, the gut must stop potentially dangerous invaders, such as bacteria and viruses, from getting inside the body. If a pathogen should cross the gut lining, immune cells in the gut wall secrete inflammatory substances that are detected by neurons in the ENS. The gut brain then either triggers diarrhoea or alerts the brain in the head, which may decide to initiate vomiting, or both.

You needn't be a gastroenterologist to be aware of these gut reactions – or indeed the more subtle feelings in your stomach that accompany emotions such as excitement, fear and stress. We now know that the ENS influences the brain. In fact, about 90 per cent of the signals passing along the vagus nerve – the super-highway that connects the brain to many of the body's organs, including the heart – come not from above, but from the ENS.

The feel-good factor

The second brain also shares many features with the first. It is made up of various types of neuron, and produces a range of hormones and around forty neurotransmitters of the same class as those found in the brain. In fact, neurons in the gut are thought to generate as much of the neurotransmitter dopamine as those in the head. Intriguingly, about 95 per cent of the neurotransmitter serotonin in the body at any time is in the ENS.

What are these neurotransmitters doing in the gut? In the brain, dopamine is a signalling molecule associated with pleasure and the reward system. It acts as a signalling molecule in the gut too, transmitting messages between neurons that coordinate the contraction of muscles in the colon, for example. Also transmitting signals in the ENS is serotonin – best known as the 'feel-good' molecule involved in preventing depression and regulating sleep, appetite and body temperature.

Does this mean the gut influences mood? Obviously the gut brain doesn't have emotions, but seems to be able to influence those that arise in your head. Nerve signals sent from the gut in response to eating fatty food, for example, make us feel good.

There is further evidence of links between the two brains in our response to stress. The feeling of 'butterflies' in your stomach is the result of blood being diverted away to your muscles as part of the fight or flight response instigated by the brain. However, stress also leads the gut to increase its production of ghrelin, a hormone that, as well as making you feel more hungry, reduces anxiety and depression. Ghrelin stimulates the release of dopamine in the brain. In our evolutionary past, the stress-busting effect of ghrelin may have been useful, as we would have needed to be calm when we ventured out in search of food. In fact, the strong links between our gut and our mental state could have evolved because a lot of information about our environment comes from our gut.

How far can comparisons between the two brains be taken? Most researchers draw the line at memory, and the things we call 'gut instinct' or 'gut reaction' originate not from the gut but the brain inside your head. And as for conscious, logical reasoning, the second brain doesn't do that. But when it comes to moods, decisions and behaviour, our second brain certainly plays a big part.

HOW GUT BACTERIA MESS WITH YOUR MIND

There are far more bacteria in your gut than cells in your body, and their weight roughly equals that of your brain. It is now becoming clear that certain gut bacteria – dubbed psychobiotics – can positively influence our mood and behaviour. These bacteria have a vast array of genes, capable of producing hundreds if not thousands of chemicals, many of which influence your brain. For example, the bacterium *Lactobacillus rhamnosus*, which is used in dairy products, has potent anti-anxiety effects in animals and has also been shown to alleviate OCD-like behaviours in mice. The hope is that one day these psychobiotics could be used to treat conditions such as depression – though we're still a long way from the development of clinically proven treatments.

WHAT IS A BRAINWAVE?

YOU'VE JUST HAD a brainwave. Oh, and there's another. And another! In fact, you will have had thousands since you started reading this sentence. These waves of electricity flow around our brains every second of the day, allowing neurons to communicate while we walk, talk, think and feel. These neural rhythms knit together everything we experience and underpin almost everything going on in our minds, including memory, attention and even our intelligence.

So, what exactly is a brainwave? Despite the way it is bandied about in everyday chitchat, the term has a specific meaning in neuroscience, referring to rhythmic changes in the electrical activity of a group of neurons. When many neurons fire at the same time, we see these changes in the form of a wave, as groups of neurons become excited, silent, then excited again, at the same time. At any one time a number of brainwaves are sweeping through the brain, each oscillating at a different frequency, classified in bands called alpha, beta, theta and gamma, each associated with a different task.

Alpha waves hum at frequencies of 8–12 Hz and are involved in our awareness and attention. Beta waves oscillate at slightly higher frequencies; they are associated with how our brain controls our muscles and how we perceive our surroundings. Lower frequency theta waves are involved with perception and memory, whereas the highest frequency waves, gamma, are to do with awareness, attention and perception, and control the flow of information stored in the brain. A newly discovered brainwave – coined the Princess Leia wave because their pattern in the brain resembles the hairstyle of the character from the *Star Wars* movies – cycles round our brain as we sleep, and may help us to remember the day's events.

You can't feel these brainwaves in action, but you will be familiar with what they look like. They're the squiggly lines picked up by electroencephalogram (EEG) readings of the brain's electrical activity. This rhythmic activity turns out to be the perfect way to organise all the information hitting our senses. Every sensation we experience, from the itch of a sweater to the buzz of a mobile phone, triggers a shower of neural signals. Brainwaves may provide clarity in this electrical storm by synchronising all the activity corresponding to a single stimulus – the words on this page, say – to a particular frequency, while neurons attending to another stimulus fire at a different frequency. This allows brain cells to tune in to the frequency corresponding to their particular task while ignoring irrelevant signals, in much the same way as we tune in to different wavelength to pick up radio stations.

Signal synchronisation

The importance of signal synchronisation becomes clear when you consider that the different aspects of a sensation – colour and shape in vision, for example – are processed in different parts of the brain before being sent to another region that binds them back together. Imagine you are looking at an apple. The apple's redness and roundness are picked up by different cells in the brain, but you don't see a red thing and a round thing – you see one item. The rhythmic activity of brainwaves ensures that all the relevant signals relating to the sensation arrive at the binding region at exactly the same time. This allows the receiving neurons to process the signals together, recombining them into a single sensation.

The specific characteristics of brainwaves, such as the timing of each wave's rhythm, influence what we see, hear and remember. For example, the strength of synchronisation between individual neurons determines how strongly we perceive certain characteristics of an image, such as its brightness. Another key role for brainwaves is dealing with memory. Low-frequency gamma waves send old memories to certain regions of the brain, whereas high-frequency gamma waves send information about what is going on at the present moment.

Retuning

Brainwaves could even explain disorders like schizophrenia. The brainwaves of people with and without schizophrenia differ; waves in those with the condition either don't spread far enough in the brain, or aren't tightly synchronised with one another. This reduced synchronisation may mean that a person with schizophrenia fails to recognise the words they have uttered as being their own, leading them to attribute the voice to someone else instead.

So could brainwaves be retuned, to help people with schizophrenia, or even to give all brains a boost? Studies have shown that volunteers hooked up to a monitor that displays an instant replay of their brainwaves have been able to suppress or activate certain brainwaves at will. This brain-training can not only boost the power of their gamma waves but also their performance on abstract reasoning intelligence tests – probably because it improves the transfer of information across the brain.

For now, we should be content in the knowledge that, for most of us, our brainwaves are working in sync. They might not make you a genius, but they are behind every thought or feeling you have ever experienced. Oh, there goes another one . . .

BRAINWAVE PASSWORDS

Watch your language. Your brain's unique response to words can reveal your identity. When you hear a particular word, your brain responds slightly differently from those of other people hearing the same word. This is because the meaning you associate with the term 'bee', for example, subtly differs from that of another person, which creates an individual electrical signal in your brain – a unique identifier that could one day be used to verify your identity as an alternative to passwords. At the moment, however, brainwave authentication is far from fail-safe as it goes awry if you are drunk, and can also be influenced by caffeine, tiredness and even a strenuous workout.

TOOLS FOR BETTER THINKING

THINKING CAN BE hard work. But thinkers don't have to work unaided. Over the centuries, philosophers have invented a range of tools to make thinking a bit easier. Some are useful only in very specific circumstances, such as calculus or probability theory. Others are more broadly applicable.

Perhaps the best known is *reductio ad absurdum* – literally, reduction of an argument to absurdity. The trick here is to take an assertion or conjecture and show that it leads to preposterous or contradictory conclusions. Homeopathy's claim that water has a 'memory' of substances that were once dissolved in it can be challenged in this way by pointing out that tap water has had millions of different substances dissolved in it. *Reductio ad absurdum* is one of the top thinking tools recommended by the philosopher Daniel Dennett, who calls it 'the crowbar of rational inquiry'.

Sturgeon's law is another of Dennett's favourites. It was named after science fiction author Ted Sturgeon, who felt that his genre was unfairly maligned by critics. 'They say "90 per cent of it is crud",' he complained. 'Well, they're right . . . but 90 per cent of everything is crud.' This is a useful tool when criticising a discipline, school of thought or art form.

If you can't land a punch on the good 10 per cent, leave it alone.

Another of Dennett's recommended tools for sharper thinking is Occam's razor: don't invent a complicated explanation for something if a simpler one will do. This is only a rule of thumb but it has proved extremely useful in science. A good example is when the heliocentric model of our solar system – in which Earth and other planets orbit the sun – swept away the elaborate system of mini orbits revolving round larger orbits, which used to explain the motion of planets around Earth.

Occam's razor is not to be confused with Occam's broom, which is the intellectually dishonest trick of ignoring facts that refute your argument in the hope that your audience won't notice. But this is just one of a number of tricks you should look out for, employed by purveyors of nonsense to give an air of credibility to their beliefs. A key one is the 'surely' tactic. Whenever you encounter this word in an argument, stop and think. The author usually wants you to skate over it, as if the claim is so obvious as to be beyond doubt, or the answer self-evident. The opposite is often the case. You should also keep an eye out for rhetorical questions. According to Dennett, a rhetorical question has a question mark at the end, but it is not meant to be answered. It represents the person's eagerness to take a short cut.

Spotting nonsense

Philosopher Stephen Law has also identified some tricks that can lead us to believe in nonsense. Watch out for the 'playing the mystery card' strategy, he says. This is when someone appeals to mystery to get them out of intellectual hot water. When defending, say, paranormal beliefs, they might say: 'But this is beyond the ability of science or reason to decide' (often followed by the quote from Shakespeare's *Hamlet*: 'There are more things in heaven and earth, Horatio,/ Than are dreamt of in your philosophy'). When you hear that, your alarm bells should go off.

Also watch out for the 'going nuclear' tactic. This can happen when someone is cornered in an argument, and they decide to get sceptical about reason. They might say 'But reason is just another faith position.' This is the 'going nuclear' tactic, because it brings every belief – that milk can make you fly or that George Bush was Elvis Presley in disguise – down to the same level so they all appear equally reasonable or unreasonable. But you can be sure, says Law, that the moment the person has left the room, they will continue to use reason to support their case if they can, and to trust their life to reason: trusting that the brakes on the car will work or that a particular medicine will cure them of an ailment.

What else can we do to become wiser? Dennett thinks that making mistakes is crucial. 'Mistakes are not just opportunities for learning; they are, in an important sense, the only opportunity for learning or making something truly new,' he writes in his book *Intuition Pumps and Other Tools for Thinking*. The chief trick to making good mistakes, he says, is not to hide them – especially not from yourself. Instead of turning away in denial when you make a mistake, you should become a connoisseur of your own mistakes, turning them over in your mind as if they were works of art, which in a way they are. The fundamental reaction to any mistake ought to be this: 'Well, I won't do that again!'

THOUGHT EXPERIMENTS

Sometimes an experiment is impossible. But that doesn't stop us from doing it – in our heads. Such thought experiments are one of the most impressive demonstrations of the power and scope of human thought – others expose weaknesses in our thinking. They have a long history.

The ancient Greeks knew about thought experiments in mathematics. Today they are most common in physics. Galileo described the first, which dealt with the speed at which stones of different sizes would fall when dropped.

The most famous is Schrödinger's cat, a creature that is both dead and alive, with its fate depending on the decay of a radioactive atom. Erwin Schrödinger devised this thought experiment to demonstrate the implausibility of a certain interpretation of quantum mechanics and was later proved right in a real-world version of the experiment.

Einstein performed another famous one at the age of sixteen, when he imagined himself running alongside a beam of light. This flight of fancy, he later said, sowed the seed for his theory of special relativity.

07.

MEMORY AND LEARNING

HOW WE RECALL
THE PAST

WITHOUT MEMORY YOU would be unable to hold a meaningful conversation or cook a simple meal. You would be eternally stuck in the here and now, unable to plan for the future. Memory is at the core of our identity and personality. Take the case of Clive Wearing – a professional musician who was unable to form memories after a brain infection. His diaries record the endless series of fits and starts that mark a life confined to a perpetual now. 'I am awake' or 'I am conscious' he wrote repeatedly, often believing that he had just awoken from a coma.

But memory is a puzzle. It can mean so many things: where you left your toothbrush last night; your mother's birthday; how to spell 'eczema'; how to play the cello part in a Beethoven quartet. And why do only certain things make it into our memory bank? The ancient Greek philosopher Plato famously compared memory to a wax tablet that is blank at birth and slowly takes the impression of events during our lifetime. Only in the past hundred years have researchers developed the techniques to study it objectively. They range from tests of our ability to remember words to more recent brain-imaging approaches.

It has become clear from these studies that, unlike Plato's wax tablet, human memory has many different components. If you consider how long a memory lasts, then there appear to be at least three different types of storage: sensory, short term and long term. Memories can also be distinguished by the type of information that is stored and the way it is recalled.

Sensory memory

During every moment of an organism's life, its eyes, ears and other sensory organs are taking in information and relaying it to the nervous system for processing. Our sensory memory store retains this information for a few moments to create a fleeting impression of a sight immediately after you experience it. Twirling a sparkler, for example, allows us to write letters in the air thanks to the fleeting impression of its path.

Sensory memories are thought to be stored as transient patterns of electrical activity in the sensory and perceptual regions of the brain. When this activity dissipates, the memory usually fades too. While they last, though, they provide a detailed representation of the entire sensory experience, from which relevant pieces of information can be extracted into short-term memory and processed further via working memory.

Short-term and working memory

When you hold a restaurant's phone number in your mind as you dial the number, you rely on your short-term memory. This temporary store of information is capable of holding roughly seven items of information for 15 to 20 seconds, though actively 'rehearsing' the information by repeating it several times can help you to retain it for longer.

Short-term memory is closely linked to working memory,

and the two terms are often used interchangeably. There is a difference, however: short-term memory refers to the passive storage and recall of information from the immediate past, whereas working memory refers to the active processes involved in manipulating this information. Your short-term memory might help you to remember what someone has just said to you, for example, but your working memory would allow you to recite it to them backwards or pick out the first letter of each word.

Long-term memory

Important or meaningful information can get transferred to the brain's long-term storage facility. Your date of birth, phone number, car registration number and your mother's maiden name are all held here. These memories can last for years, or even a lifetime, especially if the events are unusual, arousing, or associated with strong emotions. The events of 9/11 or the death of Princess Diana, for example, are very resistant to being forgotten.

We seem to store long-term memories by their meaning. If you try to recall information after a delay, for instance, you probably won't be able to reproduce the exact wording but the gist should come back fairly easily. Long-term memories can take many forms: semantic memories concern your knowledge of facts, such as Paris being the capital of France; episodic or autobiographical memories are a collection of your personal experiences, such as a particular train journey; procedural memories are a type of unconscious memory for knowing how to do things such as tying shoelaces or riding a bike. All these different kinds of long-term memories are woven into the webs of connections between brain cells.

You make use of all these types of memory, simultaneously, all the time. When you are walking down the street talking with a friend about what you did at the weekend, your brain generates numerous sensory memories of the experience – of the birds tweeting and your friend's voice. The name of the place your friend is talking about lives on in your short-term memory bank, as well as details of the conversation. Without any conscious effort, your procedural memory allows you to make the movements needed to walk, and to reach for your nose to give it a scratch. You might pull out details from your long-term memory too, perhaps a reminiscence of visiting the place being discussed. But fast forward a couple of years and it's unlikely that you will remember this conversation with your friend – unless something shocking happens, such as being robbed or witnessing a car accident.

REMINISCENCE BUMP

Which stage of your life is most vivid in your memory? When we are older, it turns out that we are more likely to remember events from the time between adolescence and early adulthood than we are from any other stage of life, before or after. Why? It could be that this 'reminiscence bump' is due to the particular emotional significance of events that occur during that period, such as meeting one's partner, getting married or becoming a parent, and events that are life-defining in other ways, such as starting work, graduating from university or backpacking around the world.

WHAT DOES A MEMORY LOOK LIKE IN THE BRAIN?

IN THE HARRY POTTER books and films, they are silver streams that can be teased from the head with the tip of a wand. In the movie *Inside Out*, they are small glowing balls, stored in racks of shelving in our minds. But what does a memory really look like? How does your brain take information from the outside world and store it for later retrieval? The answers to these questions are surprisingly elusive.

Some early seminal work on memory was carried out in the 1960s on sea slugs, creatures that grow up to a foot long and have giant nerve cells to match. These cells can be a millimetre in diameter, fifty times larger than the biggest ones found in mammalian brains. Their size makes it possible to watch what happens when a new memory forms in them. In essence, the process is due to an electrical impulse passing through a neuron, which can spark the release of chemicals that cross the gaps, or synapses, between nerve cells. This may trigger a second neuron to fire. When sea slugs learned a simple response to a stimulus, some of their synapses were strengthened. An impulse in the first neuron was now more likely to trigger the second to fire. This turns out to be the basis of memory in any animal with a nervous system.

But it still doesn't answer our question. The human brain contains around 86 billion neurons, each one connecting with around a thousand others. That gives us something like 86 trillion synapses. When I create a memory, which of those synapses are strengthened? A major step along the route to answering this question came from one of the saddest tales of modern neuroscience. In 1953, Henry Molaison had an operation to control his epilepsy that went badly wrong. His seizures originated in the hippocampi, a pair of structures either side of the brain. So the surgeon took them out. The consequences for Molaison were huge. Unable to hold a thought in his head for long, he needed care for the rest of his life. But equally profound was the impact on neuroscience – we have learned volumes from the way the surgery destroyed some of Molaison's abilities but spared others.

Molaison seemed to retain most of what he knew before the operation, suggesting that while the hippocampi are crucial for forming new memories, they are less important for storage. His short-term memory was unaffected too – he could retain information for 15 to 30 seconds but no longer. Molaison's brain damage also revealed some important subdivisions of long-term memory. He could still learn physical skills, like riding a bike, but he had problems forming new memories of things that happened to him and learning new facts.

Making a memory

The hippocampi seem to be important, then, for the sort of memories that are central to our personal and intellectual lives. But they are not the only parts of the brain required to make a memory. The cortex, the outer layer of the brain that handles our complex thoughts and sensory perceptions of the world, also plays a key role. Say that yesterday you saw a rose and stopped to inhale its fragrance. This event was processed by specific parts of your cortex at the brain's back and sides that are responsible for vision and smell. Today, if you recall the experience, those same regions will be reactivated. Similarly, the same areas of the cortex light up in scanners both when someone first sees a picture of something and when later asked to remember it.

Studies of Molaison's brain showed that a short-term memory of, say, sniffing the rose wouldn't involve your hippocampus. But if, for some reason, you created a memory that lasted more than half a minute, then connections between the relevant areas of your cortex and your hippocampus would become strengthened. Thus the hippocampus is wired up to many different parts of the cortex and helps to glue together the different aspects of a single memory. This ability helps explain one of memory's hallmarks – that recalling one aspect of an experience can bring its other features to mind unbidden. Hearing a song on the radio can remind us of the moment we first heard it, for example.

The picture that emerges is of a memory as a discrete physical entity, a spider's web of neurons firing together because of their strengthened connections, with strands reaching across different parts of the cortex and deep down to the hippocampus, the guardian of our memory bank.

To explore this process in finer detail, researchers peer into the hippocampus using electrodes. This reveals that neurons fire only on recognising a certain thing (a place, person, object or almost anything). The finding was popularised with the idea of 'Jennifer Aniston neurons', because one of the patients in these studies happened to have an electrode placed next to a neuron that fired in response to pictures of the actor. But recognising people is the job of the visual cortex, so how does that tie into neurons in the hippocampus firing?

Recognition

Different cells in the visual cortex are great at recognising the actor under different conditions – side on, with different hairstyles, lighting and so on. The hippocampal Jennifer Aniston neurons, on the other hand, don't care about what she looks like; with them it's binary – she's either there or she isn't. They even fire when her name is spoken or written. She becomes an abstract concept.

Here's how the two systems mesh: to form a lasting memory of seeing Jennifer Aniston on a particular occasion, the cortex neurons must fire up the hippocampus 'concept' neurons. The idea is that if you bumped into her while sightseeing at the Eiffel Tower, your hippocampus Jennifer Aniston neurons would start firing at the same time as your hippocampus 'Eiffel Tower neurons'. That would strengthen the connections between them, helping to create a lasting association.

So memories can be thought of as collections of concepts, which also alludes to their role as building blocks of thought. After all, how could you request a cup of tea unless you had the concept of tea, the memory of drinking it, and of liking it, stashed away?

DÉJÀ VU

Most of us have experienced the eerie familiarity of déjà vu. It's more than a feeling that you have seen or done something before: it's a disturbing sense that history is repeating, and impossibly so. This phenomenon was once thought to be due to sensory signals in the brain being out of sync, or a distortion in time perception. However, recent research suggests that the feeling arises from the brain's memory-checking system, when there's a conflict between what we've actually experienced and what we think we've experienced.

REMEMBRANCE OF THINGS TO COME

WHEN THINKING ABOUT the workings of the mind, it is easy to imagine memory as a kind of mental autobiography – the private book of you. To relive the trepidation of your first day at school, say, you simply dust off the cover and turn to the relevant pages. But there is a problem with this idea. Why are the contents of that book so unreliable? It is not simply our tendency to forget key details. We are also prone to 'remember' events that never actually took place. Such flaws are puzzling if you believe that the purpose of memory is to record your past – but they begin to make sense if it is for something else entirely.

That is exactly what memory researchers are realising. Perhaps human memory didn't evolve just so that we could remember the past, but also to allow us to imagine what might be? The idea makes intuitive sense. When you imagine yourself, say, on a beach in a forthcoming holiday, you draw on your experiences of past trips to the ocean, conjuring up a familiar scene and then filling in the details. Maybe memory provides the raw materials for these sorts of mental jaunts.

The idea increasingly makes scientific sense, too. Evidence is accumulating of an intimate mental connection between recalling the past and imagining the future. For instance, people who have lost their memories also lose their ability to imagine the future. This idea came to prominence through a patient known as K.C., who lost his memory in 1981 after suffering brain damage in an accident. K.C. had a specific problem with his episodic memory: he knew plenty of facts but was unable to remember anything about his personal past. Researchers also noticed that something else was wrong – he could not think about his future.

Researchers began to stumble across other links between memory and future thought. Another patient, D.B., whose episodic memory was wiped out after a heart attack starved his brain of oxygen, had similar problems to K.C. He knew where he worked and what sort of company it was, but he could not remember a single occasion of having been at work. He could also understand abstract concepts about the future – that global warming would be a significant problem, for example – but he could not imagine his own future.

Projection machine

Imaging studies are shedding light on this process, by watching the brain activity of people with fully functioning memories as they remember the past and imagine the future. This shows something striking and unexpected: as far as the brain is concerned, there is very little difference between the two. The brain activity of volunteers recalling or imagining a common experience such as a birthday party, a barbecue or getting lost, produced very similar patterns of activity. Intriguingly, there was no region that lit up only when remembering the past.

This suggests that our personal past and future are tightly linked in the brain. Projecting the future may not be the major function of memory, but it is one of its primary functions. From an evolutionary perspective it makes sense. It is hard to imagine how personal recall alone might be evolutionarily useful, but if remembering how cold and hungry you were last winter helps you realise the benefits of putting food away for the next one, or convinces you to plant a few of your grains instead of eating them all, you stand a much better chance of surviving than someone who cannot project themselves backward and forward in time. In fact, it is hard to imagine how civilisation could emerge from brains that cannot imagine the future.

The tight link between our past and future also sheds light on some long-standing mysteries about memory. If our capacity to remember evolved to help us imagine and shape our future, the way our memories work should reflect that function – and indeed it does. Our memories are not flawless action replays of what actually happened: chances are you do not remember what you were wearing the day before yesterday, or which cup you drank your coffee from. Yet, if you were pressed to provide details, you would almost certainly come up with something.

This seems to be how episodic memory works in general. We remember bits and pieces of our experiences and then reconstruct them to create plausible, but not necessarily accurate, accounts of what happened. Such structures make sense, if one of the main functions of memory is to shuffle scraps of the past around in novel ways to project possible futures.

It seems that every time we think about a possible future, we tear up the pages of our autobiographies and stitch together the fragments into a montage that represents the new scenario. This process is the key to foresight and ingenuity, but it comes at the cost of accuracy, as our recollections become frayed and shuffled along the way.

FALSE MEMORIES

The ease with which we form false memories is puzzling. We all have memories that are malleable and susceptible to being contaminated or supplemented in some way. This fallibility can be demonstrated in lab studies. People shown a list of words such as 'tired', 'bed', 'doze' and 'pillow' can be easily tricked into remembering that the word 'sleep' appeared on the list too, even if it did not. They do not make the same mistake with unrelated words – for instance 'butter'. This is a good demonstration of the constructive nature of our memory. We recall the gist but not exhaustive detail.

Paradoxically, people with amnesia or Alzheimer's disease often perform better in this kind of test than people with fully functioning memories. This fact seems puzzling – why would somebody with a damaged memory perform better on certain memory tests? But if our memories are designed to remember the outline of things and fill in the rest, this 'failure' makes sense. False memories are not memory deficits at all but by-products of a normal, healthy memory.

WHAT IF WE COULDN'T FORGET?

SOME THINGS IN life are best forgotten. Unfortunately for a rare group of people, forgetting is a luxury they can only dream of. They can remember every day of their adult lives in extraordinary detail. Mention any date and they are immediately transported back in time, picturing where they were, what they were doing and what made the news that day.

This amazing and rare ability comes at a price. It locks the people who have it in an exhaustive cycle of remembering – like running a movie that never stops. One woman, known as A.J., remembers every day of her life since her teens in incredible detail. Mention any date since 1980 and she is immediately transported back in time, picturing where she was. Even when she wants to, A.J. cannot forget. She describes her constant retrieval of memories as mentally exhausting. Negative memories return to haunt her again and again. So why do these people remember so much?

The root of the extraordinary ability to recall events from the past could lie in any of the stages of normal remembering. Broadly speaking, a memory is formed in three stages: first it is encoded, then stored, and later retrieved. New memories start as the temporary excitation of synapses in a network of neurons.

When you recall a memory, those same neural pathways are reactivated. The more times this happens, the more important the brain deems the memory to be and the more likely it is to be converted into a long-term memory, by forming permanent connections between the neurons. These connections are reinforced each time the memory is recalled, making it easier to retrieve. The brain contains so many potential synaptic connections that, in theory, there is no limit to the number of long-term memories that the brain can store. So why don't we remember everything?

Studying the brains of people like A.J. with highly superior autobiographical memory (HSAM) has revealed a lot about the workings of normal memory. One theory is that people with HSAM simply carry out these three stages of memory with much greater efficiency than the rest of us. But there's another, more intriguing possibility. These extraordinary memory abilities could also be explained by a failure of the strategies our brains use to help us forget the things we don't need to remember. One thing is becoming clear: having a normal healthy memory isn't just about retaining the significant stuff. Far more important is being able to forget the rest.

Why? A system that records every detail and makes that information accessible on an ongoing basis is one that will result in mass confusion. We forget because the brain has developed strategies to weed out irrelevant or out-of-date information. Efficient forgetting is a crucial part of having a fully functioning memory.

Benefits of forgetting

In fact, memory is mostly about forgetting. Tomorrow, you will probably remember a conversation you had today reasonably well, but within a week a lot of that information will be lost. Within a year, all memory of the conversation might be gone. Our brains discard most of the sensory data they receive. This is a good thing too. Our sensory organs are constantly taking in information from the outside world and without discarding the vast majority of this data, important thoughts would be lost in a sea of useless background noise.

There are several ways that we forget. Sometimes we discard information that is out of date – an old phone number or what we ate last week, for example. Since retrieving and using information solidifies it in memory, our mind gambles that the information we rarely retrieve is safe to discard.

Another way of forgetting is down to absent-mindedness where, for example, we fail to properly encode information about where we put our keys because our attention is elsewhere.

Each of these strategies has a useful purpose, preventing us from storing mundane, confusing or out-of-date memories. We want to remember our current phone number, not an old one, and where we parked the car today, not last week.

So how come some people are better at retaining unnecessary stuff than others? It might appear to be an innate ability, but it is more likely to relate to mental habits. It turns out that people with HSAM are no better at acquiring memories – they are not superior learners – but are simply better at retaining memories. This talent may be rooted in their habitual and sometimes obsessive rehearsal of past events. A 2016 study found that people with HSAM often show obsessive behaviours, similar to people with obsessive–compulsive disorder. This makes sense if you think of the neural pathways that we use to recall memories as being like garden paths. If we don't keep them organised, they become overgrown and blocked. When you habitually recall memories, then this is like keeping those pathways clear, so you are more likely to be able to retrieve that information faster at a later date.

The bottom line is that forgetting is key to having a good memory. Next time you curse your memory when you forget a name, an appointment or your own phone number just remember that your brain is trying to do you a favour.

WHY YOU CAN'T REMEMBER BEING A BABY

Few adults can remember anything from before their third birthday, a phenomenon known as childhood amnesia. You might think you remember things from this period, but these memories are often inaccurate, or false memories based on stories other people have told you.

But there are plenty of other things we do remember from those early years. In fact, it's a crucial time for learning – we start to figure out how to move and communicate, and what we like and dislike, for example. So why don't our first autobiographical memories stay with us in the same way? Blame it on the new brain cells you were making. Studies of mice suggest that there are so many neurons being born in the brain cells of young children that this interferes with the storage of long-term memories.

BOOST YOUR MEMORY

IN THE AGE of Google, with limitless information at our fingertips, it is tempting to think that a good memory is obsolete. Of course, anyone studying for exams or learning a new skill, or just trying to remember their myriad passwords, knows otherwise. In truth, many of us aspire to better recall.

So how can we pimp our capacity to remember things?

The first thing to remember is that there's a big distinction between giving the brain's hardware a boost and improving its software systems. In other words, it's relatively easy to improve your ability to remember things through tricks and techniques, but actually changing the brain's hardwiring is a different matter. Take working memory, one of the brain's most crucial front-line functions. Everything you know and remember, whether it's an event, a skill or a fascinating fact, started its journey into storage by going through your working memory. Can it be upgraded?

For a long while, working memory capacity, much like IQ, was thought to be a fixed commodity determined mainly by genetics and our early gestation period. Then some studies in the 2000s found that there might be some flex in the system after all. The research showed that the neural systems that underlie working memory seemed to grow in response to training – tasks such as memorising the positions of a series of dots on a grid.

More importantly, these studies also found that working memory training produced improvements in cognitive abilities not related to the training. For example, children who had completed these types of mental workouts showed a leap in IQ test scores of 8 per cent. Sales of brain-training programmes boomed. Since that time, however, other studies have poured cold water on these findings. Many researchers now question whether this kind of training actually makes you better at anything more than working memory tests.

The bottom line is that it seems difficult, if not impossible, to reliably improve the biological hardware involved in memory. Even worse, it is easy to damage memory through injury or the abuse of drugs or alcohol. But the good news is that there are many ways in which you can make the best possible use of your memory.

Champions of remembering

The difference between mere mortals and memory champions is more method than mental capacity. A study of the front-runners in the annual World Memory Championships didn't find any evidence that these people have particularly high IQs or differently configured brains. But, while memorising, these people did show activity in three brain regions that become active during movements and navigation tasks but are not normally active during simple memory tests. This may be connected to the fact that they use a strategy that places items to be remembered along a visualised route. This so-called 'method of loci' was invented by the ancient Greeks at least two thousand years ago and is used by almost all top memo-risers. It involves imagining a route that they know well, such as moving around their home or travelling to work, and associating the information to be learned with landmarks along that route. They can then retrieve the information later on by making the same journey in their mind and seeing the objects connected to each landmark.

How else do the champions do it? In November 2005, Chinese businessman Chao Lu became a world record holder by reciting Pi to 67,890 places. It took him a year to memorise the stream of digits and over twenty-four hours to reel them off. Like most extraordinary memorists, Chao Lu used a set of mnemonics.

Most mnemonics are based on the principles of reduction or elaboration. As the name suggests, a reduction code reduces the information to be remembered. To memorise a long list of numbers, a mnemonist might assign consonants to each number from 0 to 9, then group the stream into four-digit chunks and convert these into words by judiciously adding vowels – a mnemonic known as the phonetic system. They might then create an image for each word and weave these into a familiar journey or create a narrative in which to place memories. Later, retracing the journey or story brings back the images, which can then be decoded into the string of digits. A similar approach can help you to remember a list of random words, even the order of a pack of cards in one viewing.

This might sound complicated, but most of us will have used versions of these tricks. An acronym such as 'Roy G Biv' helps children to remember the colours of the rainbow, as does a catchy phrase like 'Richard of York gave battle in vain'. Then there is the peg word system. This involves assigning a memorable rhyming word to a number: 'one is bun', 'two is shoe', 'three is tree'. A list can then be remembered by linking each item in the sequence to each peg word, through a memorable image.

There are many other ways of increasing your chances of recall, such as actively elaborating or rehearsing information, organising it in a new way, or attempting to explain what you are studying to someone else. But don't get too carried away. While these tricks might help you to remember specific facts, they may not improve your ability to function in everyday life. Memory champions perform astounding feats and yet they can be quite forgetful in daily life, like the rest of us.

MEMORY MARVELS

Some memory champions have talents that most of us cannot emulate. A century ago, Russian journalist Solomon Shereshevsky was studied extensively for his amazing ability to remember long lists of numbers and words. This apparently required very little effort: he could recite a list of fifty numbers, forwards and backwards, after just 3 minutes of study. It turned out that as well as using mnemonics, Shereshevsky was aided by his synaesthesia – the condition where you mix up sensory information. For him, each number had a different personality – 1 was a proud, well-built man, 2 a high-spirited woman, and so on – while the sounds of other words would produce vivid colours and tastes, making them more memorable.

HOW GOOD IS YOUR MEMORY?

TEST FIVE DIFFERENT memory skills with this quiz. Make sure you have a stopwatch and blank paper handy. Take one section at a time. Once you've studied the problem for the requisite time, turn the page over and write down what you can recall. No cheating!

Language

Take four minutes to learn the following ten words from the Basque language. Then turn over this page and write them out, with their English equivalent. You get one point per correct answer.

ENGLISH	BASQUE
white	zuri
today	gaur
seven	zazpi
leg	hanka
cheese	gazta
hello	kaixo
house	etxea
bed	ohe
cold	hotza
apple	sagar

Names

Between March 1797 and April 1865, there were 15 different US presidents. You have five minutes to learn the sequence. Score one point per correct name in the correct place.

1. John Adams
2. Thomas Jefferson
3. James Madison
4. James Monroe
5. John Quincy Adams
6. Andrew Jackson
7. Martin Van Buren
8. William Henry Harrison
9. John Tyler
10. James Polk
11. Zachary Taylor
12. Millard Fillmore
13. Franklin Pierce
14. James Buchanan
15. Abraham Lincoln

Signs

You have two minutes to learn the Egyptian hieroglyphic symbols for the following letters. Score one point for each correct symbol.

Numbers

Give yourself three minutes to learn the following sequence of 20 digits. Then write down as many as you remember in the correct sequence. You get one mark for each correct answer, up until the point where you make a mistake.

Playing Cards

You have two minutes to learn the following nine playing cards, plus their position. Score one point for each card remembered, in the correct location.

Find out how your score compares on page 412.

HOW DO WE LEARN?

LEARNING IS WHAT your brain does naturally. In fact, it has been doing it every waking minute since the months before you were born. In the womb, you learned to recognise smells and music, as well as the sound of your native language and your mother's voice.

Learning is the process by which you acquire and store useful (and useless) information and skills. But what actually happens to our brains as we learn?

As the brain processes information, it makes and breaks connections, growing and strengthening the synapses that connect neurons to their neighbours, or shrinking them back. It's like building a new railway line between popular destinations, while dismantling a disused track between places that no one visits much any more. When we are in the process of learning something new, the making of new connections outweighs the breaking of old ones. Studies in rats have shown that this rewiring process can happen very quickly – within hours of learning a skill such as reaching through a hole to get a food reward. And in some parts of the brain, notably the hippocampus, the brain grows new brain cells as it learns.

But once a circuit is in place, it needs to be used if it is going to stick. This largely comes down to myelination – the process whereby a circuit that is stimulated enough times grows a coat of fatty membrane. This membrane increases conduction speed, making the circuit work more efficiently.

What, then, is the best way to learn things and retain them? The answer won't come as a huge surprise to anyone who has been to school: focus attention, engage working memory and then, a bit later, actively try to recall it. Testing yourself in this way causes your brain to strengthen the new connection. And consciously trying to link new bits of information to what you already know makes the connection more stable in the brain and less likely to waste away through underuse.

Lifelong learning

The learning process carries on for life, so why is it so much harder to learn when we reach adulthood? The good news is that there seems to be no physiological reason for the slow-down. Instead, it seems to be a lot to do with the fact that we simply spend less time learning new stuff, and when we do, we don't do it with the same potent mix of enthusiasm and attention as the average child. If the problem isn't physiological, then it might in part be down to perfectionism. Adults tend to learn a physical skill, like hitting a golf ball, by focusing on the details of the movement. Children, however, don't sweat the details, but experiment in getting the ball to go where they want. When adults learn more like kids they pick up skills much faster.

This also seems to be true for learning information. As adults we have a vast store of mental shortcuts that allow us to skip over details. We 'guesstimate' all the time, quickly and uncon-sciously using our previous experience to guide us to know what to expect when we find ourselves in many situations, such as buying a train ticket in an unfamiliar station. But we still have the capacity to learn new things in the same way as children, which suggests that if we could resist the temptation to cut corners, we would probably learn a lot more.

But where does our innate hunger to learn come from and what purpose does it serve? Once we've mastered the basics for survival – walking, talking, etc. – why do we feel the need to learn more? Why are you reading this article when you could be idly daydreaming instead?

The term 'infovore' was introduced into the scientific lexicon by neuroscientists trying to work out why humans get a kick out of learning something new. One idea is that the neural pathways through which we learn about the world tap into the same pleasure networks in the brain as are activated by drugs like heroin. So how does information give people a high? The key might be a type of chemical receptor known as a mu-opioid receptor, which is found on the surface of some brain cells. Like other opioid receptors, it is activated by heroin, morphine or naturally produced substances called endorphins, and is found in areas of the brain that mediate pain and pleasure. Mu-opioid receptors are also found in areas that process sensory information and memories. They occur in increasing numbers along the neural pathways in these areas, from the early stages where the brain processes basic things like colour to the later stages of conscious cognition.

These areas become active when the brain is trying to interpret the information it is receiving, whether it is an image of an object, or words on a page, or the song of a bird. When this happens, the endorphins that stimulate mu-opioid receptors are released, causing a feeling of pleasure. What's more, because the number of mu-opioid receptors increases further along the neural processing pathway, information that triggers the most memories and conveys the most meaning to a person causes the greatest pleasure response. It is this bonus that compels people to browse for new information.

Does the effect ever wear thin? Yes, with repetition. Reading a book for the second time is less stimulating than reading it for the first time. That is, unless you didn't understand it the first time. Endorphins are released at the 'click' of comprehension, and until the penny drops people are happy to return to a subject. Children take longer to 'click' than adults – which explains their enthusiasm for hearing the same bedtime story night after night.

WHEN IS IT TOO LATE TO BOTHER?

Even if you are well and truly past your student days, your memory can still do great things, with a bit of effort. Take the remarkable feats of a former high-school teacher known in scientific reports as JB, who started training his memory at the age of 58. He can now recite all 60,000 words of John Milton's epic poem *Paradise Lost* with amazing accuracy.

JB had shown no exceptional predisposition to memorisation beforehand, and he didn't even use mnemonics – just willpower and over 3,000 hours of daily practice. Many other people could achieve similar feats with enough time and effort. As they say, it's never too late to learn.

HARNESS THE POWER OF KNOWLEDGE

KNOWLEDGE GOES BEYOND memory, forming a rich understanding of your world. One of the brain's most useful features is the ability to absorb pieces of information and make connections between them. But what exactly is knowledge? How are facts stored, organised and recalled when needed?

Knowledge is more than just information. Even the nematode worm *Caenorhabditis elegans*, owner of one of the smallest brains we know, forages to maximise information about its environment, and therefore its chances of staying alive and reproducing. But as far as we know *C. elegans*, or indeed any species other than our own, doesn't ponder the universe's origins. Knowing, as we understand it, involves abstracting information and interpreting it for use at different times and in other contexts. Knowledge allows us to deal with entirely new situations in creative ways.

Reading this article – weighing up its truths, beliefs and justifications – won't get you a square meal or make you more attractive to a potential sexual partner (or perhaps only indirectly). And yet brain-imaging studies show that when we answer trivia questions, areas associated with our response to food and sex light up, which suggests we treat knowledge as a similar primary reward.

The precise details of how we first came to love knowledge may always elude us. But it is easy to see how it would have spurred our success as individuals and as a species, furnishing us with the tools – often literally, if you think of cutting blades or fire – to survive and prosper. So we are in some way addicted to knowledge because it has served us so well in the past – as it still does today, in everyday life as well as at the frontiers of technological progress.

Knowledge isn't so much about what information you store as how you organise it to create a rich and detailed understanding of the world that connects everything you know. It obviously relies on memory – in particular the type of memory that stores general information about objects, places, facts and people, known as semantic memory. This is the part of memory that knows Paris is the capital of France – but not the part that stores memories of a weekend break there. The sight of a dog, for example, automatically activates other bits of information about dogs: how they look, smell, sound and move, the fact that they are domesticated wolves, and your feelings about dogs.

Tagging system

How the brain achieves this gargantuan feat is far from clear. One proposal is that it has a 'hub' that tags categories to everything we know and encounter, allowing us to connect related things. A possible location for this hub is the brain's anterior temporal lobe (ATL). The ATL is badly affected in people with semantic dementia, who progressively lose their knowledge of the meanings of words and objects but retain their skills and autobiographical memories. Experiments show that, when the ATL is temporarily knocked out by an electromagnetic pulse, people lose the ability to name objects and understand the meanings of words.

Without this system we would spend a lot of time being confused about how things fit together. For example, if you were making a collage and the sticky tape ran out, how would you work out that you could use a glue stick instead? The tape is not similar to the glue stick in its shape, colour or how you use it. You need a representation that specifies similarity of kind.

So our knowledge of the world comes from a vast store of information in the brain, plus a system to retrieve and link it. But doesn't it get overloaded? It's a common trope that, as our compendium of knowledge grows, it outstrips the capacity of one brain to house it.

Limits of knowledge

The good news is that there seems to be no limit to the knowledge that can fit into a brain. As far as we know no one has ever run out of storage space. But in today's world the sheer amount of raw information to be processed undoubtedly far outstrips the capacity of any one person to process it. A human brain has roughly 86 billion neurons connected in labyrinthine ways by around 86 trillion synapses. That amounts to an information storage capacity of around a petabyte – that's a thousand times greater than the capacity of a state-of-the-art laptop with a 1 terabyte memory. Such comparisons are, of course, facile. Creating knowledge is about a lot more than assimilating data, and your brain is not an empty petabyte memory stick. If it were, you would send it back to the shop, disappointed by its slow upload rate.

And this is the rub when it comes to working out how much any individual brain can know: we have never filled one up.

We invariably reach a time limit before we reach a processing limit. Take hyper-polyglot Alexander Argüelles. Already competent in over fifty languages, he says, 'Give me total freedom of time . . . and I could conceivably do 100 languages' – at the expense of everything else in his life, though.

We shouldn't let our brains' meagre bandwidth get us down. If the amount and complexity of human knowledge has increased over time, so the means of acquiring it have improved too, with spoken and written language, the printing press and now the internet. In that profusion of information, the barrier to progress lies not in the quantity of knowledge our brains can hold, but in its quality.

CAN YOU KNOW TOO MUCH?

It is one of life's eternal mysteries: why does it get ever more difficult to recall the name of the person you were just introduced to? Surely it is a no-brainer that our cognitive powers fade as we grow older? Research seems to back this up: as we age, our scores in tests of cognitive ability decline.

Not so, according to some experts, who argue that this decline is actually down to the accumulation of a lifetime's knowledge. According to them, our cognitive skills slow down with age not because the brain withers but because it is so full. This means that it takes longer to search the larger 'mental dictionaries' that older people have built up. Learning increases the amount of information that our brains have to process, and this inevitably affects performance in cognitive tests.

DO ANIMALS
EVER FORGET?

EVERY MORNING, YOU take a walk in the park, taking some bread to feed the pigeons. As the days wear on, you begin to see the birds as individuals; you even start to name them. But what do the pigeons remember of you? Do they think kindly of you as they drop off to sleep at night, or is your face a blank, indistinguishable from the others strolling through the park? These questions may seem whimsical, but knowing what other creatures recall is crucial if we are to understand their inner lives.

If you take memory to mean any ability to store and respond to past events, even the simplest organisms meet the grade. Blobs of slime mould, for instance, which can slowly crawl across a surface, seem to note the timing of changes to their climate, slowing their movement in anticipation of an expected dry spell – even when it never actually arrives.

Driving forces

With the emergence of the first neurons about half a billion years ago, memories became more intricate as information could be stored in the patterns of electrical connections within the nervous system. Over the following few hundred million years, increasingly advanced skills could emerge with different forces driving the evolution of each creature's mind. The result is a surprising range of mnemonic feats throughout the animal kingdom.

Young chimps, for example, can beat adult humans in a task involving remembering numbers. The test entails memorising the numbers 1 to 9 appearing at random locations on a touch-screen monitor. Using an ability akin to photographic memory, the young chimps were able to memorise the location of the numerals with better accuracy than humans performing the same task (the performance of adult chimps, however, was not so slick).

Migratory cardinal fish can remember where they laid their eggs during the breeding season and, after over-wintering in deep water, return to within half a metre of the same spot. Animals as diverse as lizards, bees and octopuses can learn the way out of a maze, and pigeons have an excellent visual recognition, learning to recognise more than a thousand different images. They can even recognise individual humans and aren't fooled by a change of clothes.

Mental time travel

Such skills, although impressive, don't match our experiences of episodic memory, in which we immerse ourselves in specific events. A pigeon might learn to associate your face with food, but it probably can't remember your last meeting in the way you might be able to recall details of your last trip to the park. It is an important distinction, because episodic memory is thought to allow us to imagine and plan for the future. This skill, known as mental time travel, was long thought to be unique to humans, but there are now some signs that a handful of other species might also be able to escape the present.

Some of the most convincing evidence comes from western

scrub jays. These birds can learn from their experiences to anticipate the actions of other birds. As mentioned earlier, if a scrub jay knows that another is watching it bury food, it will later move the stash, presumably to prevent it from being stolen. But they will only do this if they have previously stolen food themselves – suggesting that they were drawing on their memories. Similar studies have suggested that bonobos and orangutans are also capable of mental time travel. And look at the behaviour of Santino, a chimp at Furuvik Zoo in Sweden that collects and hides rocks to throw at visitors. This entails premeditation, a skill that relies on episodic memory. Given the many survival benefits of being able to imagine the future, it is not surprising that other creatures show a rudimentary ability to think in this way.

Other clues about animal memories come from rats, which store mental maps of the world in their hippocampus. Studies show that different places are processed by distinct groups of neurons in the hippocampus that fire together in sequence as, for instance, when rats run around a maze. Later, after exploring an environment like this, these firing sequences have been seen replaying as the animals sleep, as if dreaming of the routes they'd taken. This process is thought to allow memories to become consolidated for longer-term storage, and has recently been detected in people for the first time. Scanning studies have also demonstrated rat brains replaying scary memories as they sleep.

But this doesn't mean that animal memories rival our own. Episodic memory depends on a host of different components, and although some animals may be able to use limited foresight when it comes to food, for instance, only humans demonstrate the kind of capacity and flexibility that allows us to imagine all kinds of futures. Santino might be able to plan a rock attack – but he could not plan anything like making a bid for freedom.

LOST IN THE HERE AND NOW

Diane Van Deren is one of the world's elite ultra runners. In one race she ran more than 1,500 kilometres over twenty-two days. On some of those days, she ran for as long as twenty hours. Van Deren had always been good at sport, but her incredible endurance seems to be down in part to her poor short-term memory, the result of brain surgery for epilepsy.

Often, she just cannot remember how long she has been running for, underestimating the time by as much as eight hours. Her inability to remember how long she has been running seems to free her from the feelings of fatigue that plague other runners. Perhaps, while others get caught up in the details of where they have been and where they are going, she gets into a more Zen-like state that lets her run for longer without feeling so much strain.

08.

THE SELF

WHO ARE YOU?

IT'S THERE WHEN we wake up and slips away when we fall asleep, maybe to reappear in our dreams. It's that feeling we have of being anchored in a body we own and control and perceive the world from within. It's the feeling of personal identity that stretches across time, from our first memories, via the here and now, to some imagined future. It's all of these tied into a coherent whole. It's our sense of self.

This intuitive sense of self is an effortless and fundamental human experience. Humans have pondered the nature of the self for millennia. Is it real or an illusion? And if real, what is it, and where do we find it? Different philosophical traditions have reached radically different conclusions. At one extreme is the Buddhist concept of 'no self', in which you are merely a fleeting collection of thoughts and sensations. At the other end is the idea that the self exists as a separate 'field' that interacts with and controls the brain. Modern science, if anything, is leaning towards Buddhism. Our sense of self is not an entity in its own right, but emerges from general-purpose processes in the brain.

Physical and psychological

Some fundamental strands underlie our idea of who we are. One is that we have a physical self – that we inhabit a body, and this is the seat of our subjective awareness. We can recognise ourselves in a mirror or photograph. A second strand is that of our psychological self. This comprises our personal traits, our memories and our unique perspective on the world. These strands are bridged by our sense of agency – an awareness of our own actions. This attributes the actions of the physical self to the psychological self, telling us that our mind is in control of the things we are doing. Neuroscientists are now able to pinpoint some of the brain processes underlying these strands. For instance, the physical sense of self seems to be centred on the brain's temporo-parietal cortex. It integrates information from your senses to create a sense of embodiment, a feeling of being located in a particular body in a particular place.

But looking at the bigger picture, it is clear that there is no particular place in the brain that gives rise to the self. In fact, if you make a list of what's needed for a sense of self, there is hardly a brain region uninvolved. A sense of self turns out to be something that emerges as the result of most parts of the brain working together.

Everywhere and nowhere

According to one model, the self is a 'nested hierarchy'. This means that the higher functions of self – self-consciousness for example – depend on the lower functions, like the basic awareness of our environment. So the higher functions of the evolutionary newest part of the brain, the cerebral cortex, require the more primitive, instinctive and emotional functions of our 'reptilian brain'. Within the brain, it seems, the self is both everywhere and nowhere.

Look closely enough, and many of our common-sense beliefs about selfhood begin to unravel. One fundamental belief is that we are unchanging and continuous. This is not to say that we always stay the same, but that today's 'me' is the same person I was a decade ago and will be in the future. Yet during our existence, we undergo great changes in our beliefs, attitudes and moods. One day we might be furious, the next contented. Yet the same self experiences both these states.

In fact, many of the things that we think define us – speaking Chinese or liking chocolate, being cheerful or even being conscious – are changeable states, yet their disappearance does not affect the fundamental essence of our self. But then it becomes unclear why such a minimal self should have the central status that we give it in our lives.

Disorders of self

Mental disorders also make it clear that this entity we regard as inviolate is not so. For example, people with schizophrenia sometimes harbour delusions that experiences and thoughts are being implanted in their brain by someone or something else. In some sense this is a disorder of the self, because these people are doing things but not feeling as if they themselves are doing them.

Even the narrative we have of ourselves as children growing up, becoming adults and growing old, which is carefully constructed from our bank of autobiographical memories, is error prone. Studies have shown that each time we recall an episode from our past, we remember the details differently, thus altering ourselves.

So the self, despite its seeming constancy and solidity, is constantly changing. The only reason we believe otherwise is because the brain does such a stellar job of pulling the wool over our eyes. Some thinkers even go so far as claiming that there is no such thing as the self. It is just an illusion.

WHAT IF THE SELF IS AN ILLUSION?

There are fewer things harder to let go of than our sense of self. Our concept of ourselves as individuals in control of our destinies underpins much of our existence, from how we live our lives to the laws of the land. The way we treat others, too, hinges largely on the assumption that they have a sense of self similar to our own.

So it is a shock to discover that this deeply felt truth could be smoke and mirrors of the highest order. But let's keep it in perspective. Much of what we take for granted about our inner lives, from visual perception to memories, is little more than an elaborate construct of the mind. The self is just another part of this illusion. And even if it is an illusion, it seems to serve us well.

THINK YOUR MIND IS FIRMLY ANCHORED IN YOUR BODY?

CLOSE YOUR EYES and ask yourself: where am I? Not geographically, but existentially. Most of the time we would say that we are inside our bodies. After all, we peer out at the world from a unique, first-person perspective within our heads – and we take it for granted.

We wouldn't be so sanguine if we knew that this feeling of inhabiting a body is something the brain is constantly constructing. But the fact that we live inside our bodies doesn't mean that our sense of self is confined to its borders. By staging experiments that manipulate the senses, we can explore how the brain draws – and redraws – the contours of where our selves reside. One of the simplest ways to see this in action is by an experiment that's now part of neuroscience folklore: the rubber hand illusion.

Sleight of (rubber) hand

The set-up is simple: a person's hand is hidden from their view by a screen while a rubber hand is placed on the table in front of them. By stroking their hand while they see the rubber hand being stroked, you can make them feel that the fake hand is theirs.

Why does this happen? The brain integrates various senses to create aspects of our bodily self. In the rubber hand illusion, the brain is processing touch, vision and proprioception – the internal sense of the relative location of our body parts. Given the conflicting information, the brain resolves it by taking ownership of the rubber hand. The implication is that the boundaries of the self sketched out by the brain can easily expand to include a foreign object. And the self's peculiar meanderings outside the body don't end there.

Trading places

A team of Swedish researchers devised an experiment to transport you out of your body and into that of a life-sized mannequin. There were cameras in the mannequin's eyes and whatever it was 'seeing' was fed into a head-mounted display worn by a volunteer. The mannequin's gaze was pointed down at its abdomen. When the researchers stroked the abdomens of both the volunteer and the mannequin at the same time, many identified with the mannequin's body as if it were their own. Scans of the brains of the volunteers showed activity in certain areas of the frontal and parietal lobes, correlating with a changing sense of body ownership.

So what's happening? Studies of macaque monkeys show us that these brain regions contain neurons that integrate vision, touch and proprioception. The idea is that in the human brain such neurons fire only when there are synchronous touches and visual sensations in the immediate space around the body, suggesting that they play a role in constructing our sense of body ownership. Upset the information the brain receives, and you can upset this feeling of body ownership. Yet while this study manipulated body ownership, the person 'inside' the mannequin still had a first-person perspective – their self was still located within a body, even if it wasn't their own. Could it be possible to wander somewhere where there is no body at all?

Into thin air

Your self can even be tricked into hovering in mid-air outside the body. In one study volunteers were asked to lie on their backs and via a headset watch a video of a person of similar appearance being stroked on the back. Meanwhile, a robotic arm installed within the bed stroked the volunteer's back in the same way.

The experience that people described was significantly more immersive than simply watching a movie of someone else's body. Volunteers felt they were floating above their own body, and a few experienced a particularly strange effect. Despite the fact that they were all lying facing upwards, some felt they were floating face down. Some reported that they were looking down at their own body from above.

When this experiment was repeated inside an MRI scanner, it showed a brain region called the temporoparietal junction (TPJ) behaving differently when people said they were drifting outside their bodies. This ties in with previous studies of brain lesions in people who reported out-of-body experiences, which also implicated the TPJ. The TPJ shares a common trait with other brain regions that researchers believe are associated with body illusions: it helps to integrate visual, tactile and proprioceptive senses with the signals from the inner ear that give us our sense of balance and spatial orientation. This provides more evidence that the brain's ability to integrate various sensory stimuli plays a key role in locating the self in the body.

Understanding how the brain performs this trick is the first step to understanding how the brain puts together our autobiographical self – the sense we have of ourselves as entities that exist from a remembered past to an imagined future. The feeling of owning and being in a body is perhaps the most basic aspect of self-consciousness, and so could be the foundation on which the more complex aspects of the self are built. The body, it seems, begets the self.

I AM THE ONE AND ONLY

Think back to your earliest memory. Now project forward to the day of your death. What you have just surveyed might be called your 'self-span', or the time when this entity you call your self exists. Either side of that, zilch.

This is a little unsettling. What is it about a mere arrangement of matter and energy that gives rise to a subjective sense of self? It must be a collective property of the neurons in your brain, which have mostly stayed with you throughout life, and which will cease to exist after you die. But why a bundle of neurons can give rise to a sense of selfhood, and whether that subjective sense can ever reside in a different bundle of neurons, may forever remain a mystery.

ARE YOU A PSYCHOPATH?

EVER WORRIED THAT you might be a psychopath? You can test how psychopathic you are in the quiz below, devised by psychologist Kevin Dutton at the University of Oxford. Note, this is not a diagnosis! And remember, if you're worried you might be psychopath then you probably aren't.

The Psychopath Test

Indicate the extent with which you agree or disagree with each of the statements below. If you strongly agree give yourself 3 points, if you agree give yourself 2 points, disagree 1 point, and strongly disagree 0 points. Then add up your total and check it against the scale to provide a rough idea of where you are on the psychopathic spectrum.

1. I rarely plan ahead. I'm a spur-of-the-moment kind of person.
2. Cheating on your partner is OK so long as you don't get caught.
3. If something better comes along it's OK to cancel a long-standing appointment.

4. Seeing an animal injured or in pain doesn't bother me in the slightest.

5. Driving fast cars, riding rollercoasters and skydiving appeal to me.

6. It doesn't matter to me if I have to step on other people to get what I want.

7. I'm very persuasive. I have a talent for getting other people to do what I want.

8. I'd be good in a dangerous job because I can make my mind up quickly.

9. I find it easy to keep it together when others are cracking under pressure.

10. If you're able to con someone, that's their problem. They deserve it.

11. Most of the time when things go wrong it's somebody else's fault, not mine.

Psychopathic spectrum score *(read more at page 412)*

0–17 LOW: You are warm and empathic with a heightened awareness of social responsibility and a strong sense of conscience.

18–22 AVERAGE: You're no shrinking violet, but no dare-devil either. You have no trouble seeing things from another person's perspective but are also no pushover.

23–29 HIGH: You are decisive, self-confident and a means-to-an-end person. For you, it's not necessarily a matter of right or wrong but of what gets the job done.

30+ VERY HIGH: Should you be worried? Not necessarily. When most people think of psychopaths, Hannibal Lecter springs to mind. But being psychopathic doesn't mean that you're a serial killer – or even that you'll break the law.

WHAT MAKES YOU
THE PERSON YOU ARE

ARE YOU THE life and soul of the party, or would you rather be at home reading a book? Do you worry about saving for retirement, or blow your earnings as soon as they hit your bank account? Each of us responds to life's events in unique ways. Where do these differences come from?

Personality is an easy concept to grasp, but a difficult one to measure. Psychologists have tried various systems over the years but most now use the 'big five' model to capture the main dimensions of your general patterns of thought and behaviour. This encapsulates personality in five traits: openness to experience, conscientiousness, extroversion, agreeableness and neuroticism (or emotional stability). These are assessed through an inventory asking you to indicate the extent you agree with statements such as 'I am the life of the party' (to assess extroversion) and 'I worry about things' (to assess neuroticism).

How you rate on any one of the big five will show up in your behaviour in a particular situation. Low scorers on the neuroticism scale, which is a measure of a person's negative emotions and ability to control them, tend to be unflappable in the face of danger. Those with a high openness score tend to be imaginative rather than practical.

A person's attitude in dealing with other people is captured by the agreeableness score: are they trusting and modest, or conceited and willing to import their ideas? The conscientiousness scale measures whether they are self-disciplined rule-followers, or prone to act before thinking.

Each trait has pros and cons. Agreeableness might make you popular, but perhaps not lead to success in the world of business. Extroverts tend to have more sexual partners and better career success. They are also more likely to end up in hospital and to get divorced.

Despite being widely accepted, there are still questions over the big five model. For a start, people's scores aren't that good an indicator of how they behave when faced with real pressures and the consequences of their actions. More fundamentally, we are coming to realise that five traits aren't enough. Many traits are not covered – anything that isn't socially desirable: aggression, alienation, cruelty, manipulativeness. This has led to calls for a sixth trait, honesty-humility, to measure a person's Machiavellian tendencies.

Another big question is how fixed these traits are. Most of us consider our personality to be an integral and unchanging part of who we are – perhaps the essence of that thing we call the self. In 1887, psychologist William James went so far as to argue that it becomes 'set like plaster' by the age of thirty. His idea stuck. But is this really the case?

Childhood temperament

There's no doubt that personality is partly genetic. What's less certain is how much is down to our genes and how much to nurture. Newborn babies don't have personalities as such, but do have characteristic ways of behaving and reacting, something psychologists call 'temperament'. This includes persistence in the face of setbacks, and 'reactivity'. Very reactive babies are shy and avoid novel situations. Temperament is often viewed as the biological basis of personality, but picking apart whether this is due to genes or environment is tricky, because both these factors interact to influence it even before birth. There's evidence that mothers who are stressed during pregnancy are more likely to have an anxious child.

Experiences in childhood also shape our personalities. Young children become more extroverted and work harder when surrounded by other kids with these traits. Parental behaviour has an impact too. If parents encourage reactive infants to be sociable and bold, they grow up to be less shy and fearful. This might help to explain why temperament doesn't always predict later ratings on big five traits. Smiley babies don't necessarily go on to be extrovert, for example. And only 25 per cent of highly reactive infants were extremely shy, anxious, timid or cautious by the age of fifteen.

By adulthood, genes seem to account for about 40 per cent of the variation in each of the big five traits – at the level of the general population, rather than for any individual. But it would be wrong to assume that genes and the environment are acting independently to influence personality. They never are. No genes have been identified that are clearly linked with any one of the five traits. In other words, genes and environment interact in complex ways to shape our personality. But where's the evidence that this process stops when we reach thirty? Well, there isn't any. In fact, once psychologists got over the intuitive appeal of the idea they began finding plenty to contradict it.

Morphing personality

The main challenge comes from studies following adults over long periods, which reveal that personality morphs with age. For instance, as we get older, we tend to become significantly more agreeable, conscientious and emotionally stable. A study of nearly four thousand people aged twenty to eighty found that personality is least stable in young adulthood, and also after about age sixty. This makes sense if changes in the environment can influence personality because young and older adulthood are periods when people tend to experience significant changes in their lives.

The extent to which environmental factors shape our dispositions over a lifetime is remarkable. A comparison of results from personality tests taken by people when they were aged fourteen and again at seventy-seven failed to find any evidence for stability in individual personality characteristics. Although psychologists continue to debate the extent to which personality is plastic, especially in adulthood, there is no doubt that it can and does change.

PERSONALITY GIVEAWAYS

Be careful of what you 'like' online. You're opening a small window on your soul. In 2015, a computer algorithm predicted personality types using nothing but what people liked on the Facebook social media site. Using data from a questionnaire filled out by 86,000 people, researchers identified their 'big five' personality traits. The results were then correlated with their Facebook likes. On the basis of between 100 to 150 'likes', the algorithm could determine someone's personality traits more accurately than their friends and family, and nearly as well as their spouse.

THE ESSENCE OF
THE TRUE SELF

IN THE 1980S, evangelical Christian Mark Pierpont travelled
the world preaching that homosexuality was a sin and promoting
ways to resist gay urges. It was a deeply personal quest. He
was wracked by the very yearnings he sought to excise from
others – a contradiction he openly acknowledged.

So here's the question: which of Pierpont's attitudes reflected
his true self? Was his message about the sinfulness of homosexuality a betrayal of his essential, gay self?

Most of us are convinced that something like a true self lurks
beneath our surface attitudes and behaviour. It might be a delusion,
but it informs how we view human beings, ourselves included.

The question of the most essential element of self has trou-
bled philosophers for centuries. In the seventeenth century,
John Locke put memory front and centre, arguing that the self
is grounded in the continuity of conscious experience. So long
as you have a memory that can stitch together experiences into
a coherent narrative, you have an enduring self.

It's an appealing idea, but modern science has given us
reasons to doubt it. People with retrograde amnesia, for example,
can lose memories from before the accident or illness that
caused it while retaining the ability to lay down new memories.

They do not feel as if their self has been wiped out and nor do their caregivers.

Intuitively, though, Locke's idea of the essence of self as being something that endures across time makes sense. If it didn't, you'd have a series of fleeting selves at best, none of which was really you. There are indications things aren't quite that simple. But take it as a starting point, and your personality would seem a prime candidate for providing that continuous sense of self – were it not for the discovery that your personality can itself change over time. So if not memory or personality, what then?

Instead of speculating about the essence of the self, psychologists and experimentally minded philosophers have a new strategy: asking people. By presenting them with various scenarios about someone changing and looking at how far they intuitively feel that the person has strayed from their true self, researchers hope to get to grips with what we regard that true self to be.

Probing personality

People can be quizzed about the hypothetical case of Jim, the victim of a serious car crash whose only hope for survival is to have his brain transplanted to a new body. In different versions of this story, post-transplant Jim remains psychologically identical or selectively loses the ability to recognise objects by sight (a condition called visual agnosia) or his autobiographical memories (amnesia).

When the transplant resulted in visual agnosia, participants viewed the change in Jim as minimal. Amnesia was seen to effect a much bigger change in his identity – in line with Locke's theory. But it was a third scenario that was regarded as having changed his self the most: when brain damage resulted in the loss of his moral conscience, so that he could no longer tell right from wrong or be moved by the suffering of others.

The same seems to be true in the real world, as shown by a survey of family members of people with one of three neuro-degenerative diseases – amyotrophic lateral sclerosis (ALS), Alzheimer's and frontotemporal dementia (FTD). ALS, the condition Stephen Hawking had, causes progressive muscle loss but leaves mental abilities intact; Alzheimer's gradually erases memories; FTD leads to changes in social and moral behaviour. Relatives of people with ALS felt the identity of their loved one had changed less than those caring for someone with Alzheimer's, but relatives of people with FTD reported seeing the greatest change.

The moral self

The upshot is that, when it comes to our perceptions of others, we see the moral self as the true self. That makes sense for us as a social species. We care about people's moral character because we want to know what they'll be like as social partners. It has even been argued that the very reason we see people as having a true self in the first place, is the importance we attach to keeping track of social behaviour.

But there's an intriguing twist to this. It seems that we see everyone's true self as not only moral but also morally good, with 'good' defined by our own moral outlook. That much was clear from studies where people were told the story of Mark Pierpont, the conflicted preacher, and similar cases, and sought their reactions. A clear pattern emerged: those with liberal values were more likely to think Pierpont's gay self was his true self, and people of a more conservative bent thought the opposite.

More generally, if someone's behaviour is good in our eyes

and accords with our values, we deem it an expression of the true self. If not, it is deemed to belong to a less fundamental, 'superficial self'. So it seems we have a pretty good handle at least on what others believe to be our true selves – even if their interpretations of the moral goodness at our core don't always tally with our own.

Whether you believe that Mark Pierpont's gay self was his true self or not might depend on your own existing moral perspective, but ultimately Pierpont made up his own mind. He renounced his life of anti-homosexual proselytising and decided he was gay after all. As Shakespeare put it, to thine own self be true.

THE FUTURE IS A FOREIGN PERSON

Ten years from now, you will still be you, right? It depends on who you ask, and when. Present You, for one, is not so sure. That much is clear from several studies revealing that we often treat our future selves like complete strangers.

In one, people were asked to make decisions about how much of a disgusting cocktail to drink. Some chose for themselves, some for the next participant and some for themselves in two weeks' time. When choosing for themselves, people opted for the smallest dose. But they went for a larger amount for another person – and for their future selves.

Brain imaging points in the same direction: thinking about your current self fires up different brain regions than does thinking about your future self, which activates the same areas of the brain as when we think about other people.

WHY YOUR SENSE OF SELF ISN'T REALLY ABOUT YOU

THE FIRST TIME a baby smiles, at around two months of age, is an intense and beautiful moment for the parents. It is perhaps the first sure sign of recognition for all their love and devotion. It might be just as momentous for the baby, representing their first step on a long road to identity and self-awareness.

Identity is often understood to be a product of memory as we try to build a narrative from the many experiences of our lives. Yet there is now a growing recognition that our sense of self may be a consequence of our relationships with others. We have a deep-seated drive to interact with each other that helps us to discover who we are. And that process starts not with the formation of a child's first memories, but from the moment they first learn to mimic their parents' smile and to respond empathically to others.

Driven by relationships

The idea that the sense of self drives, and is driven by, our relationships with others makes intuitive sense. For a start, you can't have a relationship without having a self. For me to interact with you, I have to know certain things about you, and the only way I can get at those is by knowing things about myself.

There is now evidence that this is the way the brain works. Some clues come from people with autism. Although the condition is most commonly associated with difficulties in understanding other people's non-verbal social cues, it also seems to create some problems with self-reflection: when growing up, people with autism tend to learn later how to recognise themselves in a mirror and to form fewer memories of their personal experiences. Tellingly, the same brain regions – areas of the prefrontal cortex – seem to show reduced activity when autistic people try to perform these kinds of tasks, and when they try to understand somebody else's actions. This supports the idea that the same brain mechanism underlies both types of skills.

Further support for the idea comes from the work of neuroscientist Antonio Damasio, who has found that social emotions such as admiration or compassion, which result from a focus on the behaviour of others, tend to activate the posteromedial cortices, another set of brain regions also thought to be important in constructing our sense of self.

The social 'me'

The upshot is that my own self is not so much about me, it's as much about those around me and how we relate to one another – a notion that Damasio calls 'the social me'.

This has profound implications. If a primary function of self-identity is to help us build relationships, then it follows that the nature of the self should depend on the social environment in which it develops. Researchers examining autobiographical memory have found that Chinese people's recollections are more likely to focus on moments of social or historical significance, whereas people in Europe and America focus on personal interest and achievement.

Other studies of identity, meanwhile, have found that Japanese people are more inclined to tailor descriptions of themselves depending on the situation at hand, suggesting they have a more fluid, less concrete sense of themselves than Westerners, whose accounts tend not to rely on context in this way.

Such differences may emerge at an early age. Anthropological reports indicate that the 'terrible twos' – supposedly the time when a child develops an independent will – are not as dramatic in cultures less focused on individual autonomy, which would seem to show that culture sculpts our sense of self during our earliest experiences.

These disparities in outlook and thinking imply that our very identities – 'what it is that I am' – are culturally determined. Your gender, profession, age, whether you are married or have children – all of these ways we define ourselves are really cultural artefacts.

If our sense of self arises out of relationships with others, what does this mean for our sense of consciousness, which is intimately linked to our sense of self? Here, too, a social argument has been put forward to explain why consciousness evolved. The idea is that consciousness emerged alongside other developments in brain-processing that conferred the powerful evolutionary benefit of communicating our internal thoughts to others. In order for this to happen it was necessary to generate a personalised concept of self and attribute to it a sense of awareness and of being in control of actions.

Consciousness and a sense of self therefore provide an evolutionary advantage by allowing shared communication of thoughts and ideas. This extends an individual's understanding of the world. Although your self feels deeply personal, it really might not be about you.

JUST A BUNDLE OF SENSATIONS

The problem of the self – what makes you you – has exercised philosphers and theologians for millenia. Back in the 1700s, the Scottish philosopher David Hume, looking within himself, declared he found no enduring identity, only 'a bundle of sensations'. Hume was aware that most people believe they have solid and enduring identities. Yet he asserted that human beings are 'nothing but a collection of perceptions which succeed each other with inconceivable rapidity and are in perpetual flux and movement'.

Hume's analysis implies that our sense of being continuous, coherent individuals is an illusion. We have an inbuilt tendency to see ourselves as other than we actually are. Many of today's neuroscientists concur.

WHEN THE SELF BREAKS

OUR UNIFIED SENSE of self is taken for granted. We sit comfortably inside a body we feel is ours, seeing, hearing, touching and smelling. Gloomy or happy, our feelings plainly belong to us. We own our bodily actions, whether picking up a cup of coffee or playing tennis, and we can travel back and forth in time, remembering things uniquely part of our life history and imagining our future. This self appears to us seamlessly and effortlessly as a whole. But not everyone shares this sense of unity; some rare conditions cause crucial pieces to be lost.

Body integrity identity disorder (BIID)

Imagine a relentless feeling that one of your limbs is not your own. That is the unenviable fate of people with body integrity identity disorder. They often feel it so intensely that they end up amputating the 'foreign' part. Life as an amputee is better than life with an alien limb.

The first case of BIID was reported in the eighteenth century, when a French surgeon was held at gunpoint by an Englishman who demanded that one of his legs be removed. The surgeon, against his will, performed the operation. Later, he received a handsome payment from the Englishman, with an accompanying letter of thanks for removing 'a limb which put an invincible obstacle to my happiness'.

The disorder can be viewed as a mismatch between the brain's internal map of a person's body and physical reality. Neuroimaging studies have shown that the network of brain regions responsible for creating a sense of bodily self is different in people with the condition.

Depersonalisation disorder

Many people experience brief episodes of detachment. It's the unreal, spaced-out feeling you might get while severely jet-lagged or hung-over. But for others 'depersonalisation' is an everyday part of life. *The Diagnostic and Statistical Manual of Mental Disorders IV* – psychiatry's mental health 'bible' – defines it as 'a feeling of detachment or estrangement from one's self . . . The individual may feel like an automaton or as if he or she is living in a dream or a movie. There may be a sensation of being an outside observer of one's mental processes, one's body, or parts of one's body.' There is some evidence that this state is caused by a malfunction of the body's emotion systems.

One woman with depersonalisation disorder described it as feeling like she had been dropped into her body. Rationally she knew she was real, that her memories were real and that her voice was her own, but they didn't feel like they belonged to her.

The petrified self

A crucial building block of selfhood is the autobiographical self, which allows us to recall the past, project into the future and view ourselves as unbroken entities across time. Key to this is the formation of memories of events in our lives.

Autobiographical memory formation is one of the first cognitive victims of Alzheimer's disease. This lack of new memories, along with the preservation of older ones, may be what leads to the outdated sense of self – or 'petrified self' – often seen in the early stages of the disease. It could also be what causes a lack of self-awareness of having the illness at all.

Cotard's syndrome

Of all the disturbances of the self, the eeriest and least understood is Cotard's syndrome. Symptoms of this very rare syndrome range from claims that blood or internal organs have gone missing to disavowal of the entire body and a belief that one is dead or has ceased to exist. People with the delusion – who are often severely depressed or psychotic – have been known to plan their own funerals or starve to death because they feel they no longer need to eat.

What does this all mean?

These disorders highlight how the sense of self is grounded in the body, its feelings and our brain's maps of them. Mismatches between different inputs to the brain or wrongly constructed maps can leave the brain struggling for the right answer. 'Maladies of self' arise from a brain trying to make sense of conflicting internal and external signals.

Sometimes the upshot is the strange feeling that a person's limb does not belong to them, or even that their body is separate from them. People with schizophrenia may lack agency, the sense that they are carrying out their own actions. The brain compensates by trying to make judgements about the ownership of their actions from a more external viewpoint, which may force them to feel almost outside themselves.

Many brain regions are involved in these disorders, but there is one underlying theme: the anterior insula cortex's key role in the moment-to-moment sense of self as a feeling entity. Under-activation here correlates with the fog of depersonalisation, over-activation to an ecstatic sense of being connected with others. If you delve deep enough you will see your unified self as an unnervingly fluid construction.

LOSE YOURSELF IN PSYCHEDELIA

One of the most reliable – and reversible – ways to alter your sense of self is to take psychedelic drugs such as LSD or psilocybin, the active ingredient in magic mushrooms. Alongside sensory distortions such as visual hallucinations, a common psychedelic experience is a feeling that the boundary between one's self and the rest of the world is dissolving. Research has discovered why: psilocybin causes a reduction in activity in the anterior cingulate cortex, a part of the brain thought to be involved in integrating perception and the sense of self. It was assumed that psychedelics worked by increasing brain activity; it seems the opposite is true.

WHAT IF YOU HAVE NO FREE WILL?

ARE YOU THE master of your own destiny? Did you freely choose what to eat for breakfast this morning, or to open this book? Our daily life is based on an assumption so fundamental it seems unassailable: that we have free will – we have the ability to exercise conscious control over our actions and decisions. But the more we unpick the subtle knot tying conscious experience to the brain, the shakier that assumption feels.

Tests of volition

The free will debate is an old one, but in the 1980s psychologist Benjamin Libet really stirred things up by testing free will in the lab. He asked volunteers to sit for a while and then, at their own volition, move a finger. From recordings of their brain activity, he found a signal, dubbed the readiness potential, that came before the volunteers reported they were aware of wanting to move. What appears to be a simple but freely chosen decision is determined by the preceding brain activity. It is our brains that cause that action. Our minds just come along for the ride.

This experiment has been repeated many times, but surely it is nonsense. You know perfectly well that you are in control of your actions, particularly when it's something as simple as moving your finger. Not so. There is ample evidence that our personal experience of our own actions is only a small part of the picture. Many of our physical actions are achieved by an automatic pilot in our brains – why not all of them?

Libet's experiment presents a big problem for philosophers and theologians. If we do not have conscious control this means our actions are predetermined by the genetic make-up and environmental history embodied in our brains. Are we then forced to conclude that none of us – including the most vicious criminal – is responsible for our actions? If not, is there any basis for morality?

If people lost their belief in their own free will, that would have important consequences for how we think about moral responsibility, and even how we behave. Numerous studies have shown that when people are led to reject free will they are more likely to cheat, and are also less bothered about punishing wrongdoers.

Even if free will is an illusion, it is a hard one to let go of. The 'illusion' remains as strong as ever, despite evidence to the contrary. It's like the feeling you get when you see a clever optical illusion. The lines may appear curved, but you know they're straight – you've even measured them with a ruler. Yet there is nothing you can do about the 'feeling' of them being curved. In the same way, each of us feels strongly that we control our actions.

A strong belief

What if neuroscientists were one day able to predict our every action based on brain scans – would people abandon their belief in free will? This scenario has been tested in an ingenious experiment in which participants were told of a futuristic neuroimaging technology that allows perfect prediction of decisions based on a person's brain activity, recorded by a special skullcap. They were told that a woman called Jill is fitted with this cap, for a month, to predict everything she'll do with 100 per cent accuracy, including how she'll vote in upcoming elections. Contrary to expectations, 92 per cent of participants said that Jill's voting decision was of her own free will. In another version of the story, the scientists didn't just predict which way Jill would vote – they also manipulated her choice via the skullcap. In that scenario, most participants said that Jill did not vote of her own free will.

It was easy for people to see that being manipulated negated Jill's free will, but even when her behaviour was totally predictable, people still thought she acted on her own conscious reasoning, so was responsible for her actions. This suggests that, when it comes to free will, our feelings and experience overrule any facts that neuroscientists may throw at us.

Deeply unsettling

The possibility that free will might not exist is deeply unsettling. The sense of being able to choose one course of action over another is an essential part of being human.

But this might matter less than you think. Even if we don't have free will, we act and feel as though we do. Our certainty can be shaken when we are presented with evidence to the contrary, but only briefly. Even those who deny it exists behave as if they have it. We may not have the ability to choose, but we choose to think we do.

CAN PHYSICS SAVE FREE WILL?

Neuroscience seems to pour cold water on the notion of free will. What does physics say about it? One argument is that the universe, including the bits of it that make up your brain, is entirely deterministic. The state it is in right now determines the state it will be a millisecond, a month or a million years from now. Therefore free will cannot exist.

That's not the only view. Quantum physics, our most fundamental theory of how the building blocks of the universe behave, might offer a way out. This theory says that a degree of randomness and uncertainly is built into the properties and movements of particles – including those we are composed of. Scale that up and what happens in the universe can't be entirely determined. But this doesn't solve the issue of free will. If the universe – including your brain – is fundamentally random, then how can you be said to have freely chosen to do anything?

09.
CREATIVITY

YOUR CREATIVE MIND

THERE ARE MANY things that humans can't do. We can't run like a cheetah, fly like an eagle or echolocate like a bat. What we can do better than any other species, though, is generate ideas. We come up with lots of them all the time. For better or worse, our ingenious inventions have allowed our species to take over the planet.

We tend to take our imaginative moments for granted, but creative thinking is more than just run-of-the-mill problem-solving. It requires the ability to imagine familiar objects in a different form – seeing pictures in the clouds, for example – and to make new representations of what we encounter, say, by sketching a face in the dirt. We don't know if any other creatures see the world this way, but they certainly don't spend much time representing their internal worlds in art. Humankind is almost certainly unique in its ability to imagine things that have never existed: blending what we already know into imaginary new things or finding original solutions to problems.

Ancient creations

The earliest hints of this creative talent appeared in our ancient ancestors 3.3 million years ago, when they invented the first stone tools. But it took more than 3 million years for the first symbolic objects to appear – a sure sign that early humans had a talent for abstract creations. Around 100,000 years ago, ancient humans etched a cross-hatch pattern into a block of ochre at the Blombos Cave in South Africa. Around 90,000 years ago shell beads, adorned with ochre, were created in the Qafzeh Cave in Israel. Then, starting around 50,000 years ago, there is a veritable explosion of creativity. Early humans created bone flutes, the breathtaking cave paintings of the Chauvet Cave in France, imaginative personal ornaments such as ivory beads carved to look like shells, and figurines incised with geometric patterns. Two examples that stand out are the lion-human statues from the Swabian Jura region of Germany and the painting of a bison-woman from Chauvet, both fantastical, imaginary creatures.

The ability to reproduce a three-dimensional form on a two-dimensional surface, or to 'see' a figure in ivory, requires a completely different way of imagining the world. Several key evolutionary changes had to happen to make this kind of thinking possible. Standing upright might have been one such change, narrowing the pelvis so that babies had to be born earlier, with underdeveloped brains. This led to a long period of helplessness and an extended childhood, with plenty of time for imaginative play. Other evolutionary pressures led our brains to specialise, developing dedicated mental modules that deal with specific types of thought, and allowing us to combine different types of knowledge or ways of thinking to create new ideas.

Theory of mind

Living in larger social groups also helped move things along. The need to keep track of our complex social lives led to the development of 'theory of mind': the understanding that others have thoughts and beliefs different from our own. While this skill probably evolved to help us to keep track of the group, it also allowed us to run 'thought experiments' about what others might think and do, and as a side effect to weigh up the consequences of different courses of action. The evolution of language took this to the next level, enabling us to share our thoughts and ideas with the group and develop them in many minds. The development of writing gave us another boost; now we could record ideas so that they could be worked on and improved. Around 10,000 years ago, the transition from nomadic hunter-gathering to a settled farming lifestyle brought many minds together, and a surplus of food meant less time spent foraging and more time thinking and creating.

Taken together, these changes add up to a brain that functions as an ideas generator. But one key mystery remains: where do these ideas come from? Because creative thought just seems to 'arrive', the credit has been laid at the feet of gods and spirits or, recently, the subconscious mind.

Moments of genius

Another view is that rather than actually making something new, creativity is the ability to discover things that are already in the world: that when an idea takes off, it's because it reveals a basic truth about the world that makes sense to everyone. The *Mona Lisa* may be so revered only because Leonardo da Vinci managed to capture some truth about humanity, for example, while $E=mc^2$ is considered a work of creative genius because it was there all along, waiting to be discovered by the right mind. Or, as Einstein himself put it, 'Imagination . . . is the preview of life's coming attractions.'

The challenge for understanding how these moments of

genius arise, and how to make them more likely to occur, is that we only become aware of them at the moment of insight, when the idea becomes available to consciousness. What happened before that has remained a mystery. Now, though, with the benefit of brain imaging and other innovations, we are beginning to gather some clues.

DO OTHER ANIMALS HAVE THE POWER OF IMAGINATION?

Humans are probably uniquely imaginative. But some researchers who study apes and other tool-making animals such as crows and scrub jays see powers of imagination there too. These creatures seem to be able to plan ahead when making complex tools and to solve new problems without resorting to trial and error.

Not everyone is convinced that this adds up to an active imagination. Chimpanzees and other animals may be able to conceive of something that they can't see, but whether they could weigh up two possible solutions to a problem in their mind is difficult to say. And we have no idea whether they could entertain the notion of something as far-fetched as a unicorn, or dream up imaginary worlds full of virtual friends.

WHAT GOES ON IN YOUR HEAD WHEN NEW IDEAS FORM?

UNLESS YOU BELIEVE in divine intervention, it stands to reason that ideas come from somewhere inside the brain. But there's a problem: ideas are crafted beneath the radar of consciousness, so even the greatest genius has no idea where they come from. To find out, scientists have had to develop some ingenious ways of watching them form.

The creative process

One of the earliest studies of the creative process used a network of scalp electrodes to record the pattern of brainwaves as people made up stories. This showed that creativity has two stages: inspiration and elaboration, each of which has distinct brainwave patterns, signifying very different states of mind. While people were dreaming up their stories, their brains were surprisingly quiet, mostly featuring alpha waves, indicating a very low level of arousal. This is the sign of a relaxed state, as though the conscious mind was quiet while the brain was making connections behind the scenes. It's the same sort of brain activity as in some stages of sleep, dreaming or rest, which could explain why sleep and relaxation can help people be creative.

But when these quiet-minded people were asked to work on their stories, the alpha wave activity dropped off and the brain became busier, revealing more organised ways of thinking. The people who showed the biggest difference in brain activity between the inspiration and development stages were also the ones who produced the most creative storylines, suggesting that being able to have ideas and then implement them may come down to an ability to move flexibly between these two states.

Better wiring

Activity in the brain also seems to shift during the creative process between the frontal areas behind the eyebrows and more distributed networks further back in the brain. The frontal parts of the brain are responsible for our more sensible, logical thoughts, and seem to act as a brake on our most imaginative ideas. In experiments where the prefrontal cortex was disrupted using brain stimulation, people were able to come up with more creative solutions to a mental problem.

What's more, recent studies of the brain's white matter – the brain's wiring – have found that the most creative people had less well-insulated wiring between the prefrontal cortex and other parts of the brain. Since insulation speeds up electrical signals through the brain, this suggests that a lack of it would mean slower communication, and perhaps a brain that is slow to shut down its ideas.

That's not to say that these more logical parts of the brain aren't required, however. When people are asked to list as many uses for an object as they can – the frontal lobes of their brain become noticeably more active. The frontal lobes seem to work in combination with another brain area, the anterior cingulate cortex, which monitors the brain for conflict – in other words, ideas that will never work. Together these two areas are responsible for filtering out many of the unhelpful ideas and also in helping to shift our attention away from a terrible idea towards another alternative, and so on and so on until we find one that will work.

Chemical boosters

There also seem to be chemical differences between brains that are more creative and those that are less so. The brain signalling chemical noradrenaline (also called norepinephrine) controls how easily neurons 'talk' to each other. Low levels of it appear to encourage broad networks of neurons to communicate, whereas higher levels focus that activity into tighter, smaller networks.

Giving people extra noradrenaline has been found to hinder their ability to solve word puzzles, and drugs that block the chemical can help people do better at spotting anagrams. And brain states that are characterised by the kind of low-intensity activity that favours creativity, such as sleep and depression, feature low levels of noradrenaline.

But there is more to creativity than simply having the right brain set-up and favourable chemistry. We can all learn to use what we have more effectively. Skills, situations and our social setting can shape our creativity just as dramatically as the brain resources we are born with. The most creative people also use the different rhythms of the day, the weekends and the holidays to help shift focus and brain state. They may spend two hours at their desk then go for a walk, because they know that pattern works for them.

RIGHT BRAIN, LEFT BRAIN

Pop psychology loves the idea that there is such a thing as a left-brained or right-brained type of person – the latter being the creative, intuitive type. Yet while this is, at best, a huge oversimplification, there is a kernel of truth in the idea.

Studies that track people's brain activity using EEG have found that people who solve problems by insight tend to have higher resting-state activity in the right hemisphere than more logical types. What's more, studies of people who have suffered brain damage in their left hemisphere – where much of our language processing happens – have found that it seems to bring a flood of creativity. This could be explained by language and creativity vying for processing power in the brain, and when the language centres of the left hemisphere are taken out of action creativity is more likely to break through.

THE POWER OF
IMAGINING 'WHAT IF?'

FRED LIVED INSIDE a shaggy lounge rug. Loula and Loulac were responsible for any naughty behaviour. And Charlie Ravioli was always too busy to play with the little girl who made him up. There seems to be no limit to the kind of imaginary friends that children create.

Adults spend an incredible amount of time in made-up worlds too. While most of us have ditched our childhood imaginary friends, we indulge through books, films and daydreaming instead – visiting places that only exist in our heads, inventing and exploring fictional scenarios. This habit was once regarded as irrelevant cognitive doodling. Now it is recognised as a crucial part of human thought. But why devote so much time to an activity that seems a pointless use of brainpower?

When watching children playing, it is easy to forget that this is a serious business. Playing 'pretend' helps us to rehearse the skills we need to have healthy relationships and make decisions in later life. It allows us to test out different courses of action and imagine the consequences of different behaviours without ever having to suffer them. And it seems to provide tangible benefits right from the start.

Psychologists have found that children who play pretend the most are better at seeing a problem from both sides, and imagining what would happen if circumstances change. This provides the perfect test bed for deciding how to react. A child might say, for example: 'Let's pretend we are kittens and you want to come and live in my house . . .' By running through the scene several times with a friend they are able to explore the various motivations of characters in a story and role-play their reactions to various outcomes. Will the homeowner let the kitten stay, or will she throw it out in the cold? What would happen if she threw them both out? Would the crocodiles in the garden eat them? It doesn't matter how implausible the scenario, it's all grist for the creative mill.

Imaginary friends, once considered to be a sign of a lonely kid with no social skills, seem to serve a similar purpose, but more specifically tuned to our emotional and social development. It is now thought that, far from being the preserve of the socially inept, being able to conjure up a friend from the ether may help children develop theory of mind: the ability to understand that other people see the world differently and to put yourself in their shoes. It may also give vulnerable kids a much-needed emotional lifeline.

A long-term study of children from problem families found that those who had an imaginary friend fared much better in school and later life, and had better mental health as adults than those who didn't. Some used them as a stand-in for real family and friends at a time when they felt emotionally alone. And, contrary to popular belief, the vast majority of kids are well aware that their imaginary friend isn't real.

Some people even hang on to them well into adulthood, although few admit it. Agatha Christie was still chatting to her imaginary friend at the age of seventy. While our long child-hoods are the perfect time to get lost in our own imagination, we can still dabble as adults whenever we have a quiet moment.

Wandering minds

Daydreaming is the perfect way to get back that childlike feeling. It has been shown to be crucial in boosting creativity and problem-solving, by allowing the brain to forge connections between pieces of information that we don't link up when we are too focused. Just like when we were kids, mind-wandering allows us to transcend from the present and move mentally to other places and times – as well as into the minds of other people. To imagine 'what if'.

Think about it, and you can see the everyday benefits of this kind of thinking. Imagine the following scenario: you were interviewed for a new job, but just heard that you didn't get it. You mull over all the things that happened in the interview and how they might have played out differently. Why didn't you see that question coming? How could you have forgotten to mention the brilliant idea you had prepared ahead of time? In other words, you think about a reality that didn't happen. This is how we learn from our mistakes and adjust our behaviour, and is seen as being one of the prime functions of imagination.

Similarly, every day we play out various scenarios in our minds to enable us to select the best one. Is it better to carry on working tonight, or quit now and finish off early tomorrow? Should I call my sister to tell her my news? Sometimes these spill over into pure fantasy, from daydreams about future holidays to visualisations of what a new romantic relationship might be like.

This kind of mental flight of fancy is an important part of how we make decisions. It allows us to explore our emotional reactions to various outcomes without having to actually experience them. Imagination also plays a role in designing and innovation. Every human-made object in your line of vision was imagined before it became real.

Our endless capacity for imaginative thought is what allowed us to dream up our most complex social structures, from religious organisations and civil institutions, to money, laws, and even science.

SCIENTIFIC CREATIVITY

The American physicist Luis Alvarez is widely regarded as one of the most brilliant and creative people of the twentieth century, not least because of his impact on disciplines beyond his own. Early in his career, Alvarez settled a long-standing argument among astronomers about the nature of cosmic rays: were they photons of high-energy light or charged particles?

He solved this conundrum by inventing a cosmic ray telescope that could determine the direction of the particles. He used a similar telescope to show that the Pyramid of Khafre in Giza, Egypt had no hidden chambers. And he also proved that the layer of clay laid down 65 million years ago between the Cretaceous and the Tertiary eras contained iridium that must have had an extraterrestrial origin, most likely from an asteroid impact.

Alvarez was certainly brilliant, but his productivity did not come from a scattergun approach to science. Instead, he applied a relatively small set of techniques to a wide range of areas – and this was the key to his extraordinary scientific creativity.

IS CREATIVITY LINKED TO MENTAL ILLNESS?

'THERE IS NO great genius without a tincture of madness.' So wrote the Roman philosopher Seneca, nearly two thousand years ago. It is often said that mental illness is the price to pay for a creative mind, and many creative geniuses are known or suspected to have struggled with mental ill health.

The mathematician John Nash battled schizophrenia while he developed the theory that earned him a Nobel prize, and Vincent van Gogh, Isaac Newton and Virginia Woolf have all been retrospectively diagnosed with bipolar disorder. But there's a problem with the idea that creativity and mental illness go hand in hand. Full-blown psychosis or depression are definitely not conducive to creative accomplishment. So how are the two linked?

Psychiatrists view mental health as a spectrum, with serious illness at one end and 'normality' at the other. One possibility is that a milder form of mental illness brings a creative boost that is advantageous enough to keep the genes for mental problems flowing through the gene pool. This would explain why schizophrenia continues to exist in around 1 per cent of people worldwide, despite people with the disorder having fewer children.

There is a good deal of evidence for this view. Some features of schizophrenia, including hallucinations, hearing voices, having disorganised thoughts, and believing in magic are surprisingly common, and are not a sign of mental illness in themselves. People who have some of these traits tend to score highly on tests of creative thinking.

People with schizotypal personality disorder (STPD), a condition that is similar to, but often milder than, schizophrenia, also perform particularly well on standard tests for creativity. Brain-imaging studies have thrown some light on how their brains might be wired differently. Volunteers with STPD used larger regions of their prefrontal cortex, which is responsible for thinking and other higher cognitive processes, than either people with schizophrenia or volunteers without either condition. They also had thicker connections between the two sides of the brain and more even brain activity on both sides.

It could be that greater communication across both sides of the brain allows them to throw the net wider when gathering ideas, but they are protected from full-blown mental illness by a stronger than average prefrontal cortex. The prefrontal cortex is known to keep a brake on mood and emotions, and also has a hand in narrowing down the creative possibilities that our conscious mind gets to see.

A gene called neuregulin 1, which dampens activity in the prefrontal cortex and weakens connections with the rest of the brain, seems to play an important role in all of this. A mutation for the gene has been linked to a higher risk of both STPD and schizophrenia. People with two copies of the mutation score higher on creativity tests than people with one copy, and these people in turn scored higher than people with no copies of the gene.

Too much information

Sensory overload is another feature of mental illness that has been linked with creativity. Our senses are continuously feeding a mass of information into our brains, most of which has to be blocked or ignored to save us from sensory overload. Some psychologists believe that people whose brains block out less of this information may be naturally more creative – but only if their working memory works well enough to juggle multiple sources of information at once and extract useful stuff from the noise.

Those whose brains are set up better to handle this influx of information are able to get the benefits of richer pickings from the environment to turn into great ideas. People whose brains are less well set up to cope may become overwhelmed and vulnerable to mental illness.

Bipolar disorder, which has also long been linked with creativity, may provide a different route to genius. In bipolar disorder, a person swings from depression to episodes of mania where they are euphoric, energetic and confident.

While depressive periods don't offer much in the way of creative output, manic episodes, characterised by the ability to think clearly and quickly, may prove more fruitful. Again, people with less severe manifestations of the disorder, and those who are related to people with the full-blown version are the ones that show the greatest creative benefits.

There is also some evidence that bipolar disorder might come as a downside of having a naturally fast-moving brain. In a long-term study of 700,000 students, those who got the best grades in their exams were between three and four times more likely than their less-gifted peers to develop the disorder later in life. Those whose A-grades were in creative topics such as music and literature were even more likely to suffer than those in more fact-based topics like chemistry, physics and maths.

Studies such as these will one day be able to settle the debate over whether madness and genius are flipsides of the same coin.

SEEING THE WORLD DIFFERENTLY

Everyone has their own unique window on the world, but those who are particularly creative may have a wider window than most. In experiments, volunteers were asked to look at a red image with one eye and a green image with the other eye for 2 minutes. Usually, the brain can only perceive one image at a time, so most people reported seeing the image flip between red and green. But people who scored high on 'openness' on a personality test – a trait linked to creativity – were more likely to see the two images fused into a patchwork of red and green. This suggests that when creative people come up with seemingly crazy ideas it might be because they really do see possibilities that are invisible to the rest of us.

TEST YOUR CREATIVE SPARK

CREATIVITY IS VERY difficult to measure objectively. Nevertheless, psychologists have devised a range of problems that rely on flashes of inspiration, or test your ability to think flexibly – two important facets of the creative mind.

1. Moving only one of the sticks, can you make the sum work out correctly?

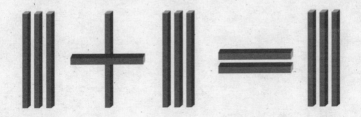

2. In two minutes, how many uses can you think up for a bucket?

3. Solve this riddle: A box without hinges, key, or lid, yet golden treasure inside is hid. What is it?

4. What word can be added to the following three to make a word or phrase?

high   book   sour

5. Move three circles to make the triangle point downwards. You can try with coins.

See answers on page 413.

HOW TO IGNITE YOUR CREATIVE SPARK

J. K. ROWLING said that the idea for Harry Potter popped into her head while she was stuck on a very delayed train. We have all had similar – although probably less lucrative – 'aha' moments, where a flash of inspiration comes along out of the blue. But is there any way to order them on demand?

Brain-imaging experiments suggest that the reason we aren't all millionaire authors is that some brains are better set up for creativity than others. EEG measurements taken while people were thinking about nothing in particular have revealed naturally higher levels of right-hemisphere activity in the temporal lobes of people who solved problems using insight rather than logic.

So far we don't know for sure whether these tendencies are born or made, but there is a good deal of evidence that everyone can get better if they use their natural abilities wisely. Thanks to a boom in creativity research over the past few years there are now a few scientifically tried and tested tips to get into the zone when you need it most.

No shortcut

The good news is that while ideas may seem to come from nowhere, in reality they spring from knowledge and information already stored in the memory. So, before it can be put together in new and exciting ways, this information needs to be stored.

Boringly, that means the first step in any creative venture is to put in the groundwork to build up information so the unconscious has something to work with. Unfortunately, there is no shortcut for this process. Studies on subliminal learning have poured cold water on the idea that knowledge can drift into the brain without any conscious effort, so it pays to focus intently on the details of the problem until all the facts are safely stored. At this stage, anything that helps with focus, such as caffeine, should help.

Once that's taken care of, it's time to cultivate a more relaxed, positive mood in which this new knowledge can be used. One option to try is to take a break and do something completely different – like watch some entertaining cat videos. Studies where people have either watched a comedy film or a thriller before coming up with new ideas have shown that a relaxed and happy mood is far more conducive to ideas than one that is tense and anxious.

A virtuous circle is set up: a positive mood brings greater productivity, and creative thoughts improve the mood: after all, there is nothing more satisfying than solving a problem. On the other hand, time pressures, financial worries and hard-earned bonus schemes have not been shown to boost creativity in the workplace. Studies suggest that internal motivation, not coercion, produces the best work.

Wide and soft focus

In general, creative ideas tend to come when the brain has a wide, soft focus, so anything you can do to turn down the dimmer switch on your attention a little will help the ideas to flow.

Perhaps the easiest way to do this is to look for ideas when your brain is too tired to concentrate properly. A 2011 study showed that morning people had their most creative ideas late at night, while night owls had theirs early in the morning. Drinking alcohol does something similar, turning down activity in the prefontal cortex, dampening not only our social inhibitions but our tendency to shoot down fledgling ideas before they have properly formed.

There are other ways to get into a more soft-focus state of mind. If you have difficulty disengaging from the problem at hand, one tip is to engage in some sensory deprivation. Brains are constantly active, so removing all external inputs leaves the mind with no other option but to wander, perhaps using the time to file memories and make sense of the day's events. This appears to make ideas easier to form and may explain why our best ideas come in the shower or while drifting off to sleep.

In experiments where people were asked to sniff a placebo that smelled of cinnamon and were told that it had been designed to enhance creativity, their performance on creative tasks improved. The researchers think it works by making them feel more confident and adventurous, and removed any fears that their ideas wouldn't be good enough. Other studies have shown that just telling people to be more creative boosts their ability, suggesting that perhaps we can all be more creative, if only we have more confidence in our abilities.

Relaxed confidence may seem to be a million miles away

when an important deadline is looming, but one day there could be an easier solution. Brain stimulation studies, in which activity was boosted in the right temporal lobe and suppressed in the left, increased the rate of problem-solving by 40 per cent. The stressed creative of the future might be able to pop on a 'thinking cap' to help those juices flow.

BORE YOUR WAY TO BRILLIANCE

According to some psychologists, our brains have an inbuilt creativity setting. All you have to do to access it is to leave your phone in your pocket and allow yourself to get bored. Experiments show that when people were made to feel bored by copying numbers out of the phone book for 15 minutes, they came up with more creative ideas about how to use a polystyrene cup than a control group who had gone straight to the cup problem. People who were asked to just read the phone book for 15 minutes were more creative still. So, if you have a problem to solve, resisting the temptation to relieve boredom might be the way to allow the mind to wander and produce flashes of genius.

10.
DECISION-MAKING

DECISIONS, DECISIONS . . .

IF I OFFERED you an all-inclusive holiday to the Bahamas, or a new house, which would you choose?

One of the key functions of the brain is to make decisions, big and small. Doing so can be difficult and perilous. Each day we face between 2,500 and 10,000 decisions, ranging from minor concerns about what coffee to drink to the question of who we should marry. But how do we make those decisions? You might believe that you use information, logic and rational thinking to make up your mind, but it's more likely that emotions, gut reactions and innate biases are the driving factors. Sometimes they push us towards a quick, accurate decision, occasionally the odd daft one.

Scientists came up with an elegant explanation for how we make choices called 'decision theory'. They called us 'rational optimisers'. When faced with a choice, they said, we weigh up each option, considering its value and probability, and then choose the one with the highest expected utility. But you might have noticed some flaws. For a start, the theory is based on the notion that humans are rational, and that we're capable of such on-the-hoof calculations of probability. In reality, we're not all-knowing beings or logical computers. We're just rather clever apes that muddle through this messy world.

Gut instincts

Actually, most of our decisions are made without any conscious effort at all – they are muddled over in the uncharted depths of our subconscious mind. But what drives these gut feelings? One idea is that we have mental rules of thumb, which allow us to make fast decisions with minimal effort. For example, we might go with a familiar option when there is little information to go on. Or we might pick the first option that meets our expectations when delaying a choice for too long is not in our interests. Researchers have shown that this second method is a sound basis for choosing a romantic partner, and that going with a familiar option is your best bet when playing a multiple choice game.

Emotions are another fundamental driver of our subconscious decision-making. Far from being the antithesis of rationality, emotions are actually evolution's satnav, directing us towards choices that have survival benefits. Anger motivates us to punish transgressors, for instance, which helps us to maintain group cohesion. Disgust, meanwhile, makes us fastidious and moralistic, which helps us to avoid disease and shun people who don't play by the rules. And while fear can lead to overreactions, this makes sense when you consider the dangers facing prehistoric humans. On that one occasion where a rustle in the bushes really was made by a predator, the less neurotic of our ancestors would have paid the ultimate price, failing to pass their laid-back genes on to the next generation.

Alternative options

Most of the decisions we make are far too complex for our brains to compute all the necessary information. Instead, we simplify things. Think about heading to your local coffee shop. Their new latte is an eye-watering £4.60. When deciding whether it is too much to pay, you might recall half-a-dozen occasions when you paid less and only two when you paid more, leading you to place this particular coffee in the 'expensive' category, and walk away.

This approach simplifies the options, but it can also lead to bad decisions when the limited information used to rank alternatives is incorrect. If, for instance, frequent nights out with boozy friends lead you to conclude that your alcohol consumption is in the top 20 per cent of drinkers when in fact it falls in the top 1 per cent you are more likely to decide to ignore the problem.

Luckily, we're clever creatures who can learn from our mistakes and favour what has worked in the past. But you still need to watch out for some hurdles – we can be swayed by our changing internal state, for example. Being hungry, thirsty, or even sexually aroused, can drastically influence our decision-making process.

Be a copycat

Sometimes, we forgo making our own decisions and simply follow the herd. In most situations it's beyond our capacity to know the best thing to do, but we're very good at recognising the people to copy. Although this can sometimes result in us replicating immoral behaviour, our conformist tendencies often lead to good choices. They allow us to fit in at a new job, or make wise purchases without knowing much about the alternatives.

A better understanding of how we make decisions could help everyone make better choices. Conformists might be persuaded to adopt environmentally sustainable habits simply because others already have. Governments wanting us to save for retirement need to understand why we are so bad at making long-term decisions. And we could all be more aware of the biases shaping our behaviour. The discovery of 'decision fatigue', which makes judges four times more likely to grant bail in the morning than in the afternoon, might persuade you to take more time out when facing a string of tough decisions. Of all the choices you face, the decision to try to make better decisions is surely the biggest no-brainer.

YOUR BAD DECISIONS

A number of mental glitches lead us to make bad decisions. We have the unfortunate habit of basing decisions on random connections. People asked to write down a high number at an auction subsequently bid more for items whose value was unknown than people who wrote down a low number.

We also overemphasise the importance of any informa-tion that confirms what we already believe. Then there's loss aversion: it feels worse to lose something than to gain the equivalent amount, making us protect what we have rather than take a chance to make a gain. When choosing whether to continue with a venture, we irrationally consider the investment we have already made in it – the sunk cost fallacy. We also love immediate rewards, which means we tend to choose smaller rewards that we can get our hands on now to bigger ones that we have to wait for.

WHY CLEVER PEOPLE MAKE STUPID DECISIONS

HOW INTELLIGENT ARE you? When it comes to making good decisions, it doesn't matter, because even the brightest people can do ridiculous things. Clever people act foolishly because intelligence is not the same thing as our capacity for rational thinking – and that's what matters when it comes to making good decisions.

IQ tests, designed to measure general intelligence, are very good at measuring certain cognitive abilities, such as logic and abstract reasoning. But they fail when it comes to measuring those abilities crucial to making good judgements in real life. That's because they don't test things such as the ability to weigh up information, or whether a person can override the intuitive cognitive biases that lead us astray. Understanding the factors that lead intelligent people to make bad decisions is shedding light on society's biggest catastrophes. More intriguingly, it may suggest ways to evade the stupidity that plagues us all.

Gut reaction

Consider this puzzle: if it takes five machines 5 minutes to make five widgets, how many minutes would it take a hundred machines to make a hundred widgets? Most people instinctively jump to the wrong answer that 'feels' right – a hundred – even if they later amend it to the correct one, which is five. When researchers put this and two similarly counter-intuitive questions to thousands of students at colleges and universities – Harvard and Princeton among them – only 17 per cent got all three right. A third of the students failed to give any correct answers.

Here's another one: Jack is looking at Anne but Anne is looking at George. Jack is married but George is not. Is a married person looking at an unmarried person? Possible answers are 'yes', 'no', or 'cannot be determined'. Most people will say it cannot be determined, simply because it is the first answer that comes to mind – but careful deduction shows the answer is 'yes' (we don't know Anne's marital status, but either way a married person would be looking at an unmarried one).

We encounter problems like these in various guises every day. And regardless of our intelligence, we often get them wrong. Why? Probably because our brains use two different systems to process information. One is deliberative and reasoned, the other is intuitive and spontaneous (see Your hare and tortoise thinking systems, page 141). Our default mechanism is to use our intuition. This often serves us well – choosing a potential partner, for example, or in situations where you've had a lot of experience. But it can also trip us up, such as when our gut reactions are swayed by cognitive biases such as stereotyping or our tendency to rely too heavily on information that confirms our own opinion. While these biases may help our thinking in certain situations, they can derail our judgement if we rely on them uncritically. For this reason, the inability to recognise or resist them is at the root of stupidity.

Think rationally

To truly understand human stupidity you need a separate test that examines our susceptibility to bias. One candidate is a test called a rationality quotient, which assesses our ability to side-step cognitive bias and work out the likelihood that certain things will happen.

So what determines whether you have a naturally high rationality quotient? More than anything, it depends on something called metacognition, which is the ability to assess the validity of your own knowledge. People with a high rationality quotient have acquired strategies that boost this self-awareness (see How to make better decisions, page 288).

But even the most rational among us can be tripped up by circumstances beyond our control. Emotional distractions are the biggest cause of error. Feelings like grief or anxiety clutter up your working memory, leaving fewer resources for assessing the world around you. To cope, you may find yourself falling back on your intuition.

Group stupidity

In the end, no one is immune to the biases that lead to stupid decisions. Yet our reverence for IQ and education means that it is easy to rest on the laurels of our qualifications and assume that we are, by definition, not stupid. That can be damaging on a personal level: regardless of IQ, people who score badly on rationality tests are more likely, for instance, to fall into debt. Large-scale stupidity is even more damaging. Business cultures that inadvertently encourage it may have contributed to the 2008 economic crisis. The effects may have been so damaging precisely because banks assumed that intelligent people act logically while at the same time rewarding rash behaviour based on intuition rather than deliberation.

Most researchers agree that, overall, the correlation between intelligence and successful decision-making is weak. The exception is when people are warned that they might be vulnerable to bias, in which case those with high IQs tend to do better. This is because while clever people don't always reason more than others, when they do reason they reason better. Which just goes to show that we should all try to be a little more aware of how we make decisions – because you are probably more stupid than you think.

HOW TO BE LESS STUPID

• Clear your mind. Judgements are often based on information you recently had in mind, even if it's irrelevant. For example, people bid higher at auctions when they are primed to ponder the height of the tallest person in the room.

• Don't fall foul of spin. We have an inclination to be influenced by the way a problem is framed. For instance, people are more likely to spend a monetary award immediately if they are told it is a bonus, compared with a rebate.

• Don't let emotions get in the way. Emotions interfere with our assessment of risk. One example is our natural reluctance to cut our losses on a falling investment because it might start rising again.

• Use facts. Don't allow your opinion to cloud your analysis.

• Look beyond the obvious.

• Don't accept the first thing that pops into your head.

AVOID THE FLAWS
THAT LEAD YOU ASTRAY

MAKING THE RIGHT decision is difficult and perilous, not least because your brain has many inbuilt biases that lead you to behave in ways that defy logic and good sense. Here's a guide to some of the choicest of these flaws. See how many you spot as you go about your day (though beware, having these biases may prevent you from being able to spot them).

Dunning-Kruger effect

The bias of illusory superiority.

This is the tendency of people with low ability to mistakenly overestimate their competence.

The Dunning–Kruger effect is a close cousin of the better-than-average effect – the statistically impossible effect in which the majority of people rate themselves more favourably than average.

There's also the reverse Dunning–Kruger effect, known as Imposter Syndrome, where a competent person feels like a fraud who is about to be found out.

Endowment effect

The tendency to value things more highly just because you own them.

'Let me pick up an ashtray from a dime-store counter, pay for it and put it in my pocket – and it becomes a special kind of ashtray, unlike any on earth, because it's mine', wrote Ayn Rand in her novel *The Fountainhead*.

This feeling is common, and leads us to make irrational decisions, like refusing to swap an item for something of higher value.

The endowment effect is one reason why the prospective purchaser of your old car won't pay the price you think it's worth.

Hyperbolic discounting

A strong preference for getting something now over something of higher value in the future.

If you were offered £50 today or £100 tomorrow, the latter would be the obvious choice. It's less clear-cut when the delay grows. Would you still wait a year for the £100? As the time gap grows, so does your preference for the immediate reward.

Hyperbolic discounting is the reason why many people's retirement funds are empty. But as retirement looms, suddenly the 'future' isn't so far away, and hyperbolic discounting comes back to bite.

Status-quo bias

A preference for the current state of affairs, and the feeling that any change from this is a loss.

This bias is linked to our desire for familiarity, and the observation that we feel more regret for bad outcomes resulting from new actions than for bad consequences that arose from inaction. It's one reason why you still drink Coke, when in blind tests you actually preferred a rival brand.

Blind-spot bias

The tendency to recognise the impact of bias in other people's judgements, but failing to see them in your own.

If you suffer from this (which you certainly do) you're not alone. Everyone thinks they are less biased than other people. This effect is related to the self-enhancement bias – the tendency to see yourself in a positive light. This blind spot means you won't be able to adapt your behaviour, even if your errors of judgement are glaringly obvious when carried out by other people.

Gambler's fallacy

The mistaken belief that if something has happened more frequently than normal, it will happen less frequently in the future.

Also known as the Monte Carlo fallacy, because of a famous incident at a casino roulette wheel there in 1913. The ball landed on black 26 times in a row and gamblers lost millions betting against black in the next spin. The chances were actually 50:50 but many people thought a red was due next time. The gambler's fallacy is the reason why, after having four boys, you think you're going to have a girl next.

HOW SUBTLE
FORCES SHAPE
YOUR CHOICES

DID YOU HEAR the one about the flies in the toilet? They took off, and started a revolution. It was 1999, and the authorities at Schiphol Airport in Amsterdam wanted to cut costs. One of the most expensive jobs was keeping the men's toilet clean. The obvious solution was to post signs reminding men not to pee on the floor. But they tried something different: they etched a picture of a fly into each urinal. The cleaning bill reportedly fell 80 per cent.

Amsterdam's urinal flies have since become the most celebrated example of a 'nudge', or strategy for changing human behaviour on the scientific understanding of what real people are like – in this case, the fact that men pee straighter if they have something to aim at. Governments across the world are increasingly employing nudges to encourage citizens to lead healthier, more responsible lives. Chances are you have been nudged, although probably without realising it.

Perfectly irrational

Back in the 1980s, economists latched on to the idea of 'rational choice theory' – that when people make choices, they exercise near-perfect rationality. Unfortunately, the theory was deeply flawed.

Imagine you are given £100 and told that you can keep it, as long as you give some to a stranger. The stranger understands the deal and can reject your offer – in which case you both get nothing. If we were purely rational creatures, we would offer a tiny amount and the stranger would accept it – even a small gain is better than none. In reality, people make surprisingly large offers, and strangers often reject ones that do not appear fair.

Real humans are not coldly rational. Although we are motivated by money, we are also motivated by social norms and the concept of fairness. Insights like this led to a new way of thinking called 'behavioural economics'. One of its most important insights is the idea that we have two systems of thought: system 1 is fast, automatic and emotional; system 2 is slow, effortful and logical. Our fast-thinking system is like our inner Homer Simpson; the slow, methodical system, our inner Mr Spock. When it comes to decision-making, system 2 generally produces better outcomes. But attention and reasoning are finite resources. So most mental tasks are left to system 1, leaving us wide open to errors.

Fish and chips

To test this, answer the following question. Fish and chips cost £2.90. A fish costs £2 more than the chips. How much do chips cost? System 1 instantly shouts out 90p. It takes deliberation to arrive at the correct answer, which is 45p.

Many other biases and flaws are at play in our decision-making. We often follow the herd instead of making choices to suit ourselves; we tend to choose the path of least resistance; we value short-term pleasure more than long-term success. But this is where things get clever: human behaviour is irrational, but predictably so. It is this predictability that convinced economists that it should be possible to change behaviour. And so the concept of nudge was born. The main tool of nudging is simply tweaking how options are presented. Supermarkets are experts at this. They greet you with the smell of baking bread and place the most profitable brands at eye level. The intention is that you cave in to temptation and buy things you didn't intend to.

A gentle push

Public authorities have woken up to the power of nudging, and how it might be used to persuade people to do the right thing. The 'right thing' is a value judgement, but is usually defined as the option people would have chosen if they were not burdened by biases.

In practice, nudging can mean all sorts of things. Many nudges simply reverse a default option, such as automatically registering citizens as organ donors. You can opt out, but most people do not get around to it. Similarly, people can be nudged into being more public-spirited by applying social pressure. People who owe unpaid tax in the UK receive a letter telling them (truthfully) that most of their neighbours pay their taxes on time. This nudge has increased compliance from 68 to 83 per cent.

Importantly, nudges do not use orthodox economic incentives like fines and rewards. Increasing the price of alcohol, for instance, might reduce drinking, but that isn't a nudge. A nudge would be prompting pubs to sell beer in smaller glasses as well as pints, on the understanding that if you give people a big amount they will probably consume it even if they don't really want to. Nudges must allow people to be free to make the wrong choice. You can still drink pints if you want to and nobody will tell you that you can't.

Will all this lead to better societies? Advocates are adamant that science is on their side. But certain nudges can prove counterproductive. For example, there is evidence that when foods are labelled as low fat, it is taken by the consumer as a licence to consume more. Perhaps the most serious obstacle to the nudge revolution is public acceptability. Although nudges are intended to be helpful and preserve freedom, many people feel there is something sinister about them. And so the question is not 'Do you want to be nudged?', but 'Who do you trust to do it?'

SUCCESSFUL NUDGES

• **Slow down:** The number of car accidents on the twisty Lake Shore Drive, Chicago, was reduced by painting white stripes on the approaches to sharp turns. The stripes get progressively closer together, creating the illusion of speeding up.

• **Eat your veg:** Researchers in New Mexico doubled the amount of fruit and vegetables people bought at a super-market by dividing shopping baskets in half and marking one of the sections 'fruit and vegetables'.

• **Save energy:** People can be encouraged to reduce electricity use by leaflets informing them about the energy-saving measures their neighbours are taking.

HOW DO WE DECIDE RIGHT FROM WRONG?

WOULD YOU CLEAN your toilet with your country's flag? Or have sex with a chicken bought from the supermarket? Like most people, you probably think these acts are immoral – despite not necessarily being able to provide a reason why.

Recently, scientists have been getting to grips with our moral thinking and behaviour – and in the process they've discovered how we might manipulate them in order to solve some of the world's most challenging problems. Humanity's ideas about morality evolved over tens of thousands of years. Early humans were forced to hunt and forage collectively or starve. This in turn fuelled the evolution of cognitive abilities underlying collective action – the ability to share goals, responsibilities and rewards.

So how do we make moral decisions today? One theory is that moral judgements are driven not by rational, reflective thought but by gut feelings, fuelled by emotional reactions. Most people are repulsed by the thought of having sex with a deceased chicken, and that alone is enough to condemn the act. When reasoning comes into play, it frequently rationalises these intuitive decisions after the event.

It certainly doesn't follow that we ought to do what our

instincts prompt us to do. That might have enhanced our survival and reproductive fitness in an earlier period, but may not do so now; even if it did, it could still be the wrong thing to do. Gut reactions may guide our judgements, but we can also stop and think, and try to make better decisions.

Default setting

Our intuitive moral sentiments are akin to the automatic settings on a digital camera, while our rational deliberations are analogous to manual mode, where we adjust everything by hand. Our automatic settings are efficient but not very flexible, whereas manual mode is flexible, but takes time to change the settings. In the same way that many of us rely on automatic settings on our camera because it is easier, we tend to make quick-fire moral judgements based on gut reactions. This kind of reaction is perfect for solving smaller problems, but not global issues like climate change or poverty.

Take empathy, for example. This aspect of our automatic mode functions like a spotlight, throwing into relief the plight of whoever falls under its beam, and moving us to action. You might think it is a force for good – but you'd be wrong. Empathy illuminates the suffering of a single person rather than the fate of millions, and it is more concerned with the here and now than the future. It's because of empathy that we care more about, say, the plight of a little girl trapped in a well than we do about potentially billions of people suffering from climate change. All this sounds a bit disheartening, but there is hope: we can shape our moral intuitions.

Morality pills

In a recent study, researchers asked people how they would feel about performing a mercy killing on a terminally ill man by various methods, including giving him a poison pill, suffocating him and shooting him in the face. You might expect opposition to each method would be predicted solely by the amount of suffering it causes. But in fact it was better predicted by participants' aversion to performing the action. It suggests that we base our instinctive moral judgements not only on our emotional reaction to suffering, but also on how the physical acts involved make us feel. And here's where we might be able to make some changes.

One way is to deliberately seek out particular experiences. Imagine an aspiring vegetarian. If bacon sandwiches are proving too much of a temptation, they might watch videos documenting the bad treatment of animals. Next time they see meat in front of them they may find it disgusting rather than appetising. The same tactics might help people avoid doing things that increase their carbon footprint, say, or add to the plight of the world's poorest.

Then there's the power of shame. Take BankTrack, a network of NGOs that exposes banks involved with projects that threaten the environment and human rights. BankTrack has compiled a list of the top 'climate killers'. Through naming and shaming, they hope to provoke a race to the top, where banks compete to clean up their portfolios and stop financing investments that are destroying the planet.

One day we might even help solve global conundrums by enhancing our brains through biomedical means. For instance, citalopram, a drug used to treat depression, can make people more sensitive to the possibility of inflicting harm on others.

One study showed that people who took citalopram were willing to pay twice as much money as controls to prevent a stranger from receiving an electric shock. Likewise, delivering electrical signals through the scalp – to stimulate brain regions implicated in regulating social emotions – can reduce stereotypical attitudes towards members of different social groups.

These effects may be subtle, but the fact that it is possible raises the prospect that someone might use it. Which provokes moral questions. Who would we treat, how and when? Should we add drugs to the water supply? Should we fortify children's cereal with moral enhancers? What would you do?

DOG'S DINNER — AND OTHER MORAL PROBLEMS

Try these moral dilemmas to probe your feelings about right and wrong:

• A family's dog is killed by a car. They have heard that dog meat is delicious, so they cook it and eat it for dinner. Is this wrong? Why?

• Julie and Mark are a brother and sister who decide to make love. Julie takes birth-control pills, but Mark uses a condom too, just to be safe. Was it wrong for them to have sex?

• I have some fresh orange juice, into which I have dipped a sterilised cockroach. The cockroach was bought from a lab and raised in a clean environment. Just to be certain, I sterilise it again so that no germs can survive. Would you drink the juice?

HOW TO MAKE
BETTER DECISIONS

DECISIONS, DECISIONS! OUR lives are full of them. We jealously guard our right to choose, it's the very definition of free will. Yet sometimes we make bad decisions that leave us unhappy or full of regret. Can science help? Here's a guide to making up your mind.

Don't fear the consequences

Whether it's choosing between a weekend in Paris or a trip to the ski slopes, almost every decision we make entails predicting the future. In each case we imagine how the outcome of our choice will make us feel.

Sensibly, we usually plump for the option we think will make us the happiest. The only problem is that we are not very good at it. People routinely overestimate the impact of decisions, both good and bad. We tend to think that winning the lottery will make us happier than it actually will, or that life would be unbearable if we were to lose the use of our legs. The real outcomes are usually less intense and briefer than people imagine.

A primary factor leading us to make bad predictions is 'loss aversion' – the belief that a loss will hurt more than a corresponding gain will please. In reality, we're very good at coping with losses, and finding new ways to see the world that make it a better place for us to live in.

So how do we avoid being a poor forecaster? Rather than imagining how a given outcome might make you feel, find someone who has made the same choice, and see how they felt. Remember also that whatever the future holds, it will probably hurt or please you less than you imagine.

Consider your emotions

You might think that emotions are the enemy of decision-making, but in fact they are integral to it. Our most basic emotions evolved to enable us to make rapid and unconscious choices in situations that threaten our survival.

Whenever you make up your mind, your limbic system – the brain's emotional centre – is active. People with damage to the emotional parts of their brains are often crippled by indecision, unable to make even the most basic choices, such as what to wear or eat. This could be because our brains store emotional memories of past choices, which we use to inform present decisions.

Whether emotions always allow us to make the right decisions is another matter. Take anger: researchers induced anger in a group of people by getting them to write an essay recalling an experience that made them see red. They were then presented with a choice: either take a guaranteed $15 payout, or gamble for more with the prospect of gaining nothing.

The researchers found that men, but not women, gambled more when they were feeling angry. In another experiment, angry people were less generous and more likely to opt for the first thing they were offered rather than considering other alternatives.

All emotions – disgust, happiness, guilt – affect our thinking, so it might be best to avoid making important decisions under their influence. Yet strangely there is one emotion that seems to help us make good choices. When making a decision, sad people take time to consider the various alternatives on offer, and end up making the best choices.

Play the devil's advocate

Ever had an argument with someone about a vexatious issue such as immigration or the death penalty and been frustrated because they only drew on evidence that supported their opinions and conveniently ignored anything to the contrary? This is the ubiquitous confirmation bias. It can be infuriating in others, but we are all susceptible. The only way to protect yourself from confirmation bias is to search for evidence that could prove you wrong. It's a painful process and requires self-discipline. But if you want to make good choices, you need to do more than pick the facts that support the option you already think is the best.

Keep your eye on the ball

Our decisions have a habit of becoming attached to arbitrary facts and figures. In a classic study that introduced this so-called 'anchoring effect', participants were asked to spin a 'wheel of fortune' with numbers ranging from 0 to 100, and afterwards to estimate what percentage of United Nations countries were African. Unknown to the subjects, the wheel was rigged to stop at either 10 or 65. Although this had nothing to do with the subsequent question, the effect was dramatic. On average, participants presented with a 10 on the wheel gave an estimate of 25 per cent, while the figure for those presented with a 65 was 45 per cent. They had taken their cue from the spin of a wheel.

Anchoring kicks in whenever we are required to make a decision based on limited information. With little to go on, we are prone to latch on to irrelevancies and let them sway our judgement. One strategy is to create your own counterbalancing anchors, but it's a hard effect to shake.

Don't cry over spilt milk

Does this sound familiar? You are at an expensive restaurant, the food is fantastic, but you've eaten so much you are starting to feel queasy. You know you should leave your dessert, but you feel compelled to polish it off despite a growing sense of nausea. Or perhaps at the back of your wardrobe lurks an ill-fitting item of clothing. But you cannot bring yourself to throw it away because you spent a fortune on it.

The force behind both these bad decisions is called the sunk cost fallacy. The more we invest in something, the more commitment we feel towards it. To avoid letting sunk cost influence your decision-making, always remind yourself that the past is the past and what's spent is spent. If when thinking about whether or not to end a project you realise that you would not start it now if you had the choice, then it's probably not a good idea to continue.

Look at it another way

Consider this: your hometown faces an outbreak of a disease that will kill 600 people if nothing is done. To combat it you can choose either programme A, which will save 200 people, or programme B, which has a one in three chance of saving 600 people but also a two in three chance of saving nobody. Which do you choose?

Now look at this situation. You are faced with the same disease and the same number of fatalities, but this time programme A will result in the certain death of 400 people, whereas programme B has a one in three chance of zero deaths and a two in three chance of 600 deaths.

Both situations are the same, and in terms of probability the outcome is identical. Yet most people instinctively go for A in the first scenario and B in the second. It is a classic case of the 'framing effect', in which the choices we make are irrationally coloured by the way the alternatives are presented. In particular, we have a strong bias towards options that seem to involve gains, and an aversion to ones that seem to involve losses. That's why healthy snacks tend to be marketed as '90 per cent fat free' rather than '10 per cent fat'.

Can we avoid framing effects? Yes, simply look at your options from more than one angle.

Beware of social pressure

You may think of yourself as a single-minded individual and not at all influenced by others, but no one is immune to social pressure. In one classic study known as the Milgram experiment, researchers persuaded volunteers to administer electric shocks to someone behind a screen. It was a set-up, but the subjects didn't know that and on the researchers' insistence many continued upping the voltage until the recipient was apparently unconscious.

Numerous studies have shown that groups of like-minded individuals tend to talk themselves into extreme positions, and that groups of peers are more likely to choose risky options than people acting alone.

How can you avoid the malign influence of social pressure?

First, if you suspect you are making a choice because you think it is what your boss would want, think again. If you are a member of a group, never assume that the group knows best, and try playing the contrarian.

Finally, beware of situations in which you feel you have little individual responsibility – that is when you are most likely to make irresponsible choices.

Limit your options

You probably think that more choice is better than less – Starbucks certainly does – but consider these findings. People offered too many alternative ways to invest for their retirement become less likely to invest at all; and people get more pleasure from choosing a chocolate from a selection of five than when they pick the same sweet from a selection of thirty.

The problem is that greater choice usually comes at a price. It makes greater demands on your information-processing skills, and it can be confusing, time-consuming and at worst lead to paralysis: you spend so much time weighing up the alternatives that you end up doing nothing. In addition, more choice also increases the chances of you making a mistake, so you can end up feeling less satisfied with your choice because of a niggling fear that you missed a better opportunity.

To counter this effect, you could try going with the first option that meets your preset threshold of requirement. This 'good enough' method takes a lot of the pressure off and makes the task of choosing more manageable. Even when 'good enough' is not objectively the best choice, it may be the one that makes you happiest.

So instead of exhaustively trawling through the websites in search of your ideal digital camera or barbecue, try asking a friend if they are happy with theirs. If they are, it will probably do for you too.

Swayed by our own opinions

Four cards, each with a letter on one side and a number on the other, are laid out. You can see D, A, 2 and 5. Your task is to turn over the cards that will allow you to decide if the following statement is true: 'If there is a D on one side, there is a 5 on the other.'

Typically, most people pick the D and 5, reasoning that if these have a 5 and a D respectively on their flip sides, this confirms the rule. But look again. Although you are required to prove that if there is a D on one side, there is a 5 on the other, the statement says nothing about what letters might be on the reverse of a 5. So the 5 card is irrelevant. Instead of trying to confirm the theory, the way to test it is to try to disprove it. The statement is false, and the correct answer is to turn over D (if the reverse isn't 5, the statement is false) and 2 (if there's a D on the other side, the statement is false).

WHEN TO GO WITH YOUR GUT

It is tempting to think that to make good decisions you need time to systematically weigh up all the pros and cons. But sometimes a snap judgement or instinctive choice is the best basis to make a decision, especially when faced with information overload.

In one study participants were asked to choose one of four hypothetical cars, based either on a simple list of four specifications such as mileage and legroom, or a longer list

---

of 12 such features. Some participants then got a few minutes to think about the alternatives before making their decision, while others had to spend that time solving anagrams.

The result? When faced with a simple choice, subjects picked better cars if they could think things through. When confronted by the complex decision, however, they became bamboozled and actually made the best choices when they did not consciously analyse the options. The researchers found that the same was true in real-world situations with complex options.

11.

THE SOCIAL BRAIN

READING OTHER PEOPLE'S MINDS

IT'S A WEEKDAY evening and two workmates, Sally and Anne, are having a drink in the pub. While Sally is in the bathroom, Anne decides to buy another round of drinks, but she notices that Sally has left her phone on the table. To keep it safe, Anne puts the phone in Sally's coat pocket before heading to the bar. When Sally returns, where will she expect to see her phone? This is a no-brainer. She will obviously look on the table, because she doesn't know that Anne has moved it. Even though you know, you also know that Sally doesn't.

This ability to put yourself in somebody else's brain is called 'theory of mind' and most people do it without a moment's thought.

Because it comes so naturally we tend to take it for granted. But it involves doing something no other animal can do to the same extent: temporarily casting aside our own ideas and beliefs about the world and taking up somebody else's. This process, also known as 'mentalising', not only lets us see the world from somebody else's perspective and predict their behaviour, it also allows us to tell lies and spot deceit by others. Until somebody invents telepathy, it's the closest we can get to mind-reading.

Your brain is doing it all the time. When you get dressed in the morning, for example, you are intuiting how people will respond to your appearance. Driving, cycling or walking to work, you are constantly predicting what other road users know about your intentions, and vice versa. Our work, social and family lives demand that we keep track of others' mental states: who knows what about whom, thinks what about whom, and what they may do about it. Imagine navigating the complex terrain of allies, rivals and relatives without being able to at least hazard a guess about what they know and think.

Even during relaxation you are often busy mentalising. TV, film and novels would make little sense without it. Compelling drama often depends upon keeping track of who knows what about whom – or at least what they *think* they know. Some have even proposed that Shakespeare's genius was to make his audience work at the edge of their ability, tracking multiple mind states simultaneously.

No theory of mind

If you want to know what a human without a theory of mind looks like, try a child of around three. Their lack of mentalising skill is best revealed by the 'Sally-Anne test', which involves acting out a version of the workmates-in-a-pub scenario, only with puppets instead of people and a ball instead of a phone. If asked 'When Sally returns, where will she look for the ball?', most three-year-olds say she'll look in the new spot, where Anne has put it. The child knows the ball's location and cannot conceive that Sally does not. But around the age of four, children make a dramatic cognitive leap. Most four- and five-year-olds realise that Sally will expect the ball to be where she left it.

Despite its importance in human relations, not everybody can do theory of mind to the same extent. Age is a key determinant. After its sudden appearance in childhood, it develops as we get older and is not fully functional until we reach our twenties. Even then theory of mind is not an all-or-nothing quality. Some people are more adroit than others in social situations, with a seemingly telepathic ability to understand what other people are thinking, and wanting and intending.

To understand the variation requires going beyond the simple Sally-Anne test, which nearly everyone over the age of five can pass like falling off a log. That scenario represents the simplest level, called 'second-order' theory of mind. But imagine introducing a third character to the scene, a would-be thief who sees Anne move the phone. Most of us have no trouble reading his mind: he wants to steal the phone but understands he must steal it before either Sally or Anne return. This is third-order theory of mind.

Fourth-order theory of mind is also straightforward for almost everybody: add in an undercover policeman who witnesses the whole scene. It is fairly easy to step into his shoes and understand what he knows about the knowledge and intentions of the other three, and also to guess what he will do about it.

But beyond that point some people start to lose track. Add in a fifth character, a mischief-maker who sees Anne looking for the phone and tells her he saw the thief take it. What is everyone thinking now? There are other complications too. Think back to Sally looking for her phone, but imagine where she might look once she realises it is not on the table. If she applies her theory of mind to her conscientious colleague Anne, she might well guess that it is in Anne's pocket.

About one in five people can't go any further than fourth-order theory of mind. A further one in five get stuck at fifth-order. Only the top 20 per cent can reach the heights of sixth order. But even if you're not in the upper echelon, it is possible to improve with practice. One of the best ways is to read literary fiction, or even watch soap operas with multiple characters and convoluted plots. Don't think of it as slobbing in front of the TV, but a serious bout of brain-training.

BIGGER GROUP, BIGGER BRAIN?

The need to keep track of convoluted social relations may have been a key driver of human brain evolution. Among monkeys and apes, those living in bigger groups have a larger prefrontal cortex, the outermost section of the brain where higher thought happens. This also applies to humans. Hunter-gatherers typically live in groups of around 150, which is the largest group size of any primate – and we have the brains to match. On an individual level too there is a correlation between the size of people's cortices, the size of their social networks and their ability to pass theory-of-mind tests.

WHY WE'RE WIRED
TO PERSUADE

FOR ANYONE WHO thinks facts and evidence matter in politics, Britain's EU referendum campaign of 2016 was a depressing spectacle.

Both sides told lies – that the economy would tank immediately if those wishing to leave the EU won, that 76 million Turks were about to become EU citizens, that leaving would free up £350 million a week to give to the National Health Service. These claims were debunked repeatedly, but campaigners kept on repeating them anyway, secure in the knowledge that telling the truth is less persuasive than telling a good story.

It is not the first time that politicians have been economical with the truth in pursuit of votes, and it won't be the last. But for a species that prides itself on its powers of rational thought, the fact that so many people can be swayed by demonstrably false claims is something of a mystery.

Emotions are part of the answer, of course: the Leave campaign was better at telling uplifting stories than the dry and technocratic Remainers.

Even so, if you think that human intelligence evolved to apply rigorous logic to complex problems, then our irrational nature is difficult to explain. But according to an alternative view of what our brainpower is for, it makes perfect sense. Our reasoning powers evolved not to find the truth, but to persuade others to accept our point of view, even if it is false.

Go with the group

The roots of this new concept of human rationality come from research into the rigours of group living. It has long been recognised that being a social species is fraught with mental challenges.

Our ancestors would have to form and maintain alliances, track who owes what to whom, and watch out for being misled by others in the group. If that were not hard enough, they would also have faced problems that could only be solved by collective action – and that meant taking collective decisions. When the right course is unclear, who should you follow? A headstrong and persuasive leader is one possibility. Democratic deliberation is another. Either way the ability to argue convincingly would have been in our ancestors' best interests.

From an evolutionary perspective that makes sense. Groups that made good collective decisions would have survived better than those that did not. And thus, the argument goes, we evolved to persuade and became skilled at building convincing arguments, not necessarily based on strong evidence.

That explains a lot about the human condition. Consider confirmation bias – the universal tendency to believe and use evidence that supports our existing position while ignoring or dismissing anything that contradicts it. Such a bias looks like a bug if we evolved to solve problems: you are not going to get the best solution by considering only one side of the argument.

But if we evolved to argue our corners, then confirmation bias takes on a much more functional role. You won't waste time searching out evidence that doesn't support your case, and you'll focus on evidence that does.

The lure of attraction

There is a similar explanation for another bias, the 'attraction effect'. When faced with a choice between different options, irrelevant alternatives can sway our judgement from the logical choice.

It is perhaps best illustrated by considering a range of smartphone contracts: people who would tend to choose the cheapest option can be persuaded to opt for a slightly up-market model if an even more expensive, flashy model is added to the mix. It seems that the luxury option sways us by offering an easy justification for the decision – we can claim that we have landed a bargain.

Notably, the attraction effect is strongest when people are told that they will have to defend whatever choice they make publicly. There's also the framing effect, whereby the choices we make are irrationally coloured by the way the alternatives are presented.

It is perhaps because of these biases that decisions taken collectively are often superior. When given a difficult logical task to solve, groups of five or six generally do better than individuals.

Groups that argue more and allow everyone to chip in are the most likely to succeed, as well as avoiding the problems of groupthink. Such results are exactly what you might expect from a species that evolved not to think individually, but collectively. It really is true that two heads are better than one.

THE CHARISMA FACTOR

Nelson Mandela. Steve Jobs. Michelle Obama. Some people just seem to have a natural ability to influence others; they've got charisma. It's one of our most prized personal qualities, important for uniting groups.

Although we know it when we see it, charisma is a rather slippery concept. To the ancient Greeks, it was *ethos*, meaning persuasive appeal. In dictionaries, it still carries magical connotations as a 'divinely conferred power or talent' that makes one individual capable of influencing or inspiring others. The field of psychology is producing its own answers. Based on decades of research, psychologist Ronald Riggio has identified six traits or skills that he believes are essential: emotional expressiveness, enthusiasm, eloquence, self-confidence, vision and responsiveness to others. To be perceived as charismatic, it is vital to have a balance between these components, he says. A surfeit of emotional expressiveness, for instance, can detract from personal charisma – just think of the comic persona of actor Jim Carrey.

THE BENEFITS
OF COPYING

SYLVIA IS A good cook. She makes a particularly fine roast ham, using a family recipe that begins, unusually, by cutting a slice off both ends of the meat. One day, a friend dropped by while she was preparing the ham. 'Why do you do that?' he asked. 'Because that's the way my mother does it,' Sylvia replied. But it got her wondering – why?

A few days later Sylvia was at her mother's house. 'When you make that ham, why do you start by chopping the ends off?' she asked. 'Because that's the way my mother taught me,' came the response. So Sylvia picked up the phone to her grandma and asked the same question. The old woman paused for a moment in recollection. 'I think it was because my roasting dish wasn't big enough,' she replied.

The story of Sylvia's family recipe illustrates what is so unusual about the way humans learn. We have a tendency to copy the actions of others without question, especially when we are young. At first glance such unquestioning imitation seems foolish because it opens us up to doing pointless things. But actually it turns out to be a good recipe for success. Our brains are uniquely adapted to copy, and this apparent weakness is among our greatest strengths.

Pointless imitation

If you want to see pointless imitation in action, try this experiment. Avail yourself of a small child, ideally aged between one and two, and a torch. Show the child how to turn the torch on and off using your nose. A week later present them with the torch again. Chances are they will try to turn it on with their nose, even though it would be easier to do it with their hands. In a formal version of this experiment done in the 1980s using a light box rather than a torch and forehead rather than nose, two-thirds of fourteen-month-olds switched the box on the hard way.

Other experiments on more complex tasks show that young children copy every single step performed by a demonstrator, even ones that are clearly irrelevant. If you could do the same experiment with a chimp, you'd see something very different. Observation of primates in the wild shows that instead of faithful imitation – copying both the means and the ends of a task – they fixate on the ends. They observe the result of the behaviour they have been watching – cracking a nut, for example – and then they figure out their own way to achieve it.

At first glance, the chimps appear to be taking the more fruitful approach, avoiding the chimp equivalent of pointlessly chopping off the ends of the ham. But there is a downside. Individual chimpanzees essentially have to discover the art of nut cracking for themselves and never pass on the useful knowledge to others to adapt or improve. So instead of incremental improvements to their technology akin to those that have taken us to the moon, chimpanzees still sit on the ground cracking nuts with stones, as they have for millions of years.

Hard-wired to copy

But faithful imitation also has its downsides. You may find yourself doing things that are pointless. What you really need is someone to show you how to perform a task, rather than just letting you watch what they do. So human brains are not just hard-wired to copy, but also to actively pass on knowledge and skills. This combination of copying and demonstrating is called pedagogy, and as far as we know it is uniquely human.

Some animals do display rudimentary forms of teaching, such as when adult meerkats disable the stinger on a scorpion to allow their offspring to learn to kill it without risk of being killed themselves. But only humans practise the systematic teaching of complex actions.

The pedagogy instinct is a two-way street. Babies are innately motivated to soak up information from adults, and parents are driven to share information. They use words, gestures and eye contact to gain and hold the child's attention while demonstrating how to do something. In the light-box experiment, children who are actively guided to learn the forehead technique are much more likely to reproduce it than children who are passively shown it.

A different path

So what sent humans down a different learning path from other apes? It could be down to the sheer volume and complexity of material we have to learn. Because of our huge cultural repertoire, children need to learn quickly how to do many things. They can modify techniques later if need be.

Humans also make and do things with no obvious immediate function. Some researchers trace the origins of pedagogy back to when our ancestors started producing complex tools, such as handles hafted onto flint axes. The mental gymnastics required to make tools, often requiring other tools in their manufacture, would have necessitated a new way of passing on information. When fashioning a handle for an axe, for instance, it would be difficult for a naive onlooker to know what the thing was for. So the only way our young ancestors could learn from their elders was to imitate their actions while being guided by their teachers. Human see, human do.

CHIMP VS HUMAN

Another difference between the way humans and chimps learn is the instinct to put heads together. In one experiment, groups of human children and chimpanzees were presented with a box with three locked chambers, each containing rewards. The chambers had to be opened in the correct order to unlock the goodies. The children used information sharing to crack the problem, showing each other how they had solved parts of the problem and splitting the rewards. Chimps went it alone, plugged away, but failed.

MORAL DILEMMAS

TO EXPLORE OUR moral intuitions, philosophers have devised a series of thought experiments called trolley problems. They highlight the dilemmas we face when deciding right from wrong.

The basic trolley problem

A runaway tram trolley is heading down the tracks towards five trapped workers, who will all be killed if the train follows its course. The only way to save them is to divert the train onto the second track with only one person on it, who will be killed. Should you divert the train, killing one person but saving five others?

This dilemma shines a light on two different moral philosophies. According to the utilitarian school of thought, you should divert the train and kill one man because this achieves the greatest good for the greatest number of people. But according to the so-called deontological moral philosophy, deliberately causing harm (even if it results in a good outcome for others) is always wrong.

The fat man scenario

A runaway tram trolley is heading down the tracks towards five trapped workers, who will all be killed. You are on a footbridge over the tracks. Next to you is a very large man. The only way to save the workers is to push a heavy object in front of the trolley. The only available heavy object is the man next to you. Should you push the stranger to save the five workers?

The fat villain

This is a variation of the fat man scenario. In this case, you know that the large man on the bridge has caused the train to come hurtling down the tracks to kill the five workers. Should you push the fat villain onto the tracks to save five workers?

FEELING LONELY? YOU'RE NOT ON YOUR OWN

IMAGINE YOU ARE a zookeeper and it's your job to design an enclosure for humans. What single feature would best ensure the health and well-being of the animals in your care? Good food? A comfy bed? A gym?

The thought experiment has only one answer. Above all else, the enclosure must take into account our need for connection with other people.

Humans are what biologists call an 'obligatorily gregarious species'. Our brains crave human contact and do not function well without it. Chronic isolation puts us at risk of a long list of neural and behavioural problems: anxiety, hostility, social withdrawal, broken sleep, depression, eating disorders and increased risk of dementia.

The problems are not just psychological. Left unchecked, loneliness can take a physical toll as bad as smoking or obesity. Lonely people are at high risk of just about every major chronic illness going, from heart attacks to cancer. All told, loneliness increases the odds of early mortality by 26 per cent.

Safety in numbers

To understand why loneliness is so unhealthy, it helps to think of it in terms of our evolved psychology. Social primates like us live in groups as a means of protection. Being separated from the group is risky, and the feeling of loneliness is a spur to make us seek out safety in numbers: think of it as a biological warning sign a bit like hunger, thirst or pain.

Similar things are seen in other social species. Fish threatened by predators prefer to swim to the middle of their shoal where they are less likely to be picked off, mice housed in social isolation suffer from sleep disruption, and prairie voles isolated from their partners explore their surroundings less and concentrate on predator evasion.

Such observations point to a general principle: social isolation in social animals activates neural, neuroendocrine and behavioural responses that promote short-term self-preservation.

In the short term, a pang of loneliness is no more harmful than being hungry. But being starved of company for too long can have serious consequences.

One reason for this is that loneliness lowers willpower, making us more likely to indulge in harmful behaviours such as overeating or failing to exercise. There are other, more subtle, effects too. One is on the immune system. Chronically lonely people have generally suppressed immune function, though one branch of the system, inflammation, is ramped up sky-high. Inflammation is the body's first line of defence against injury and bacterial infection, but too much for too long has been linked to cancer, depression, Alzheimer's disease and obesity.

Why does the immune system respond in that way? Activating inflammation while dampening other immune functions is called the 'conserved transcriptional response to adversity'. It is triggered by the fight-or-flight response and shifts resources away from our default immune function, which tends to protect us against viral infection, towards a more effective response to bacterial infection. These are the kind of microbes that tend to follow the wounding injuries you might get if attacked by a predator. Once that acute mortal terror has passed, we would normally shift back into antiviral mode. But chronically lonely people's bodies are constantly primed for an attack that never comes.

Too much inflammation for too long also changes the brain, triggering behaviours that prime us for threats. It makes us a little more suspicious, vigilant and irritable. And it is a vicious circle. The changes to the way we think and behave are not conducive to sociability. Loneliness also makes us worse at reading social situations, which just makes matters worse. The heightened sense of threat lonely people feel also means they are more likely to pay attention to and remember negative details and events, and behave in ways that confirm their negative expectations, perpetuating the vicious spiral of loneliness.

On top of that, inflammation also dampens down brain areas involved in motivating you to interact with others. This probably evolved in part as a form of self-quarantine for sick people. But in the modern world, these kinds of behaviours create a vicious cycle towards increased loneliness. Loneliness begets loneliness.

It can happen to anyone. Loneliness is assumed to be a problem of social isolation, predominantly affecting older or vulnerable people with no friends and family, who rarely leave home.

Living alone

Yet loneliness can have very little to do with being alone or with having few friends. It is about quality rather than quantity of relationships. Healthy, well-adjusted people tend to have four or five intense friendships or family relationships, which occupy about half of their total social effort. The best way to inoculate yourself against loneliness is to see them on a regular basis.

Yet this is often not easy. More people live alone, and the number of single-parent households is rising. Work leads many of us to live far away from our families. At the same time, technology has changed the way we work, shop, socialise and entertain ourselves, largely serving to reduce the amount of face-to-face contact we get.

So what do you do if it takes hold? The best approach is to start with your mind, rather than your social networks. Loneliness is a psychological problem marked by chronic feelings of threat and hostility. The best way to break through these is cognitive behavioural training. It would be an exaggeration to say that loneliness is all in the mind, but the solution to it probably is.

SOLITARY CONFINEMENT

Perhaps the most extreme example of human loneliness is seen in prisoners kept in solitary confinement. Decades of research into their psychology has found most suffer significant mental health problems. Just a few weeks in isolation can trigger panic attacks, anxiety, loss of control, irrational anger, paranoia, hallucinations, obsessive thoughts, depression, insomnia, cognitive dysfunction and self-mutilation.

Psychologists say the pathological effects stem from the lack of social interaction, without which inmates find it hard to maintain a sense of their own identity, the appropriateness of their emotions and how they relate to a wider social world. In other words, extreme loneliness theatens not just their health but their basic sense of being.

HIVE MINDS: THE POWER OF GROUP PSYCHOLOGY

IMAGINE YOU ARE given a piece of card with a line drawn on it. In front of you are three other pieces of card with lines on them, one of which is exactly the same length as yours. Your job is to pick the one that matches.

Easy. But now imagine you are put into a group of strangers and asked to do the task collectively. The others all pick a line that seems blatantly wrong to you. Would you stick to your guns or accept the judgement of the others?

When psychologist Solomon Asch tried this out in the 1950s, using plants who were told to pick the wrong line, he found that an astonishing 70 per cent of people caved in to peer pressure. Asch concluded that, for most of the volunteers, conforming to the group was more important than the evidence of their own senses.

Welcome to the powerful and unsettling world of group psychology, where individuals lose their own minds and mob rule rules.

Group mentality

Such situations are surprisingly common. You find them in just about any environment in which individuals are part of a group or are reacting to what others are doing: committees, social networks, riots, football crowds, even panels of judges. In such situations, a group mentality can easily take over, leading people to act out of character or endorse positions they would never normally adopt. Perhaps the most famous demonstration of group psychology was the 1971 Stanford Prison Experiment, where students were recruited to play the roles of guards and inmates in a mocked-up prison. After six days, the experiment had to be stopped because the guards – ordinary young men chosen for their healthy psychological state – had pushed many of the prisoners to the brink of a breakdown. In a similar experiment a few years later, Stanley Milgram of Yale University persuaded ordinary people to administer electric shocks to a 'victim' in another room. Without much trouble Milgram goaded many of them into increasing the voltage until the victim was screaming. Two-thirds of them carried on until the victim was unconscious. The shocks were not real and the victims were actors, but the participants didn't know that.

The key to both experiments was to subsume personal accountability to the needs of the group. Milgram did this by telling the participants that he was in charge and would take the rap for anything that happened. Philip Zimbardo, who ran the prison experiment, let his guards hide behind collective symbols of power – uniforms, whistles, handcuffs, sunglasses.

'If you can diffuse responsibility so people don't feel accountable, they will probably do things they normally never would,' Zimbardo later said. An analysis of 25,000 social psychology studies concluded that he was right: almost everyone is capable of evil acts if placed in the wrong context.

Social cascades

All of this is a long way from the situations that most of us face. Yet many of the decisions we make every day are heavily influenced by group psychology. For example, the reason chart-topping pop songs are more popular than average is not because they are significantly better but because consumers are influenced by the buying habits of others. This is known as the social cascade effect, a phenomenon in which large numbers of people end up doing or thinking something on the basis of what others have done.

There are two mechanisms at work here. The first is social learning. The world is too complicated for each individual to solve problems on their own, so we assume other people know things we don't. The other is social coordination, where you do the same thing as other people for the sake of group harmony. These two forces can influence financial markets, protest movements and even how we vote.

It is not surprising that people should be so easily led. After all, we evolved as social animals in environments where cooperation and group cohesion were key survival tools. But in the modern world, it often backfires. One common problem is polarisation, which is when a group of like-minded people end up taking a more extreme position than any of its individual members. For example, a group who begin a discussion believing Brexit is probably a bad idea may end up concluding that it is insanity. There are two reasons for this. First, in like-minded groups you tend to hear only arguments that support your own viewpoint, which is bound to reinforce it. In addition, people are always comparing themselves with others and will shift their position so as not to appear out of line.

The power of others

Another form of group psychology is groupthink, where members strive for cohesion at the expense of all else. Maintaining cohesion can give a group a sense of power and bolster its members' self-esteem, but it can also lead them to make bad decisions. Groupthink has been blamed for the CIA's flawed plan to invade Cuba in 1961 and also for NASA's failure in 2003 to recognise that the damage done to the wing of the space shuttle Columbia by a piece of foam was potentially fatal.

Despite the commonness of group psychology, we tend to ignore it. Society tends to focus on individual psychology and our institutions are based on this concept. Yet if we don't understand the power of the group we can never hope to combat evils such as torture, suicide bombings and genocide, or just avoid making bad decisions of our own.

The good news is that people can and do resist. Zimbardo himself went on to study heroism, which is almost the mirror image of mob rule. He found that just as people can act out of character under the influence of a group, they can also heroically break ranks in favour of doing the right thing. Mob rule doesn't always rule.

THE BROTHERHOOD

One of the most powerful forms of group psychology is the 'brotherhood mentality' that builds up – or is actively fostered – in combat. Groups of soldiers who collectively experience extreme fear often fuse into a collective prepared to do anything for each other, even if it means certain death.

The brotherhood becomes more important than the cause or country they are fighting for. This is used to extreme effect in terror cells, where carrying out a suicide mission comes to be seen as the ultimate act of group solidarity.

12.

SLEEP AND DREAMING

WHAT IS SLEEP?

WHETHER YOU THINK of it as a waste of time or it's your favourite thing in the world, sleep overcomes us all eventually. Since we spend about a third of our lives in the land of nod you'd think that scientists might have a pretty good handle on what sleep actually is. Not so.

Part of the problem is that, by definition, we're unconscious while we're doing it and remember little about it when we wake up. Worse, a sleeping body gives little away to the outside observer: the hallmarks of sleep are lying down, moving very little and failing to respond to the outside world. So, it's hardly surprising that it was long thought to be a switched-off dormant state, not worthy of serious scientific attention.

In 1953, though, a young PhD student called Eugene Aserinsky wondered if there might be more to it. He did what no one had thought of doing before: spent hours staring at the eyelids of someone who was fast asleep. He saw what we now know as REM sleep, the state in which we do most of our dreaming and which, it later emerged, features the kind of intense brain activity that you would see if someone was awake.

In the decades since then we have worked out much about what happens to the brain while we sleep. We haven't exactly nailed down all the details, though – researchers are still restricted to interpreting brain scans for the tell-tale signals of activity, or waking people up to ask them what they are experiencing.

The science of shut-eye

What we now know is that the brain is as busy during sleep as during your waking hours, perhaps even more so. And that getting enough of each sleep stage seems to be hugely important for our mental and physical well-being.

In the late 1960s, scientists studying sleeping people's brains via electrodes on the outside of their scalps (by electroencephalography, or EEG) had discovered the four stages of sleep that correspond to different brain wave patterns. We move through them in turn in a cycle that lasts for about 1.5 hours.

A full night's sleep consists of five or six cycles. After lying awake for ten minutes or so we enter non-rapid eye movement sleep or NREM sleep. NREM sleep is divided into three stages, NREM1, NREM2 and NREM3, based on subtle differences in EEG patterns. Each stage is considered progressively 'deeper'. After the NREM stages we enter REM sleep. The EEG signal during REM sleep is similar to wakefulness or drowsiness. It is during this stage that many of our dreams occur.

REM was long considered to be the most important sleep phase because after sleep deprivation people catch up the next time they sleep at the expense of other stages. Some scientists now think that slow-wave sleep is the most important phase as it is not distributed evenly through the night, but crammed into the first half, suggesting it is a priority. It's likely that each stage has its own important role to play in giving us a good night's rest.

Feeling sleepy

Our slide into slumber doesn't just happen like flipping a switch. Instead, there is a two-pronged system of brain and hormonal activity that dictates how sleepy we feel and when. First there is the body's circadian rhythm, which controls the time of day that we feel the need to sleep. The main controller of this is a tiny patch of brain tissue called the suprachiasmatic nucleus (SCN), which sits just above the optic nerve. This acts as a master clock, gathering information about light from the retina and relaying it to the rest of the body via nerve impulses and hormones. The best known of these, melatonin, rises as the SCN detects that light is fading. Using smartphone screens or other sources of blue light after dark delays melatonin release and can make it more difficult to drop off.

The pressure to drop off

Then there's sleep drive, also known as sleep pressure, the brain's measure of how long you have been up. The longer you stay awake, the more a chemical called adenosine – a by-product of metabolism – builds up in your brain. Adenosine is thought to suppress the neurons that usually keep us awake, so as levels increase, the desire to sleep builds. After sixteen hours of being awake there is so much adenosine on board that you can't help but drop off.

The dual nature of the circadian clock and sleep pressure system explains why jet lag and shift work are so disruptive. When the two processes get out of sync, we feel sleepy at the wrong time according to the circadian clock.

There is a way to cheat the system. Caffeine temporarily keeps you perky by blocking adenosine receptors in the brain, but the effects of overriding the sleep drive eventually break through. Being awake for twenty-four hours will leave you with the same level of cognitive impairment as having a blood alcohol content of 0.1 per cent – more than the drink-drive limit in several countries.

The only way to reverse these effects is to catch up on the hours of sleep you have lost. Or, if there's not enough time, a nap can go a long way.

AWAKE, ASLEEP OR NEITHER?

It was once thought that you could be in one of two states: asleep or awake. But we now know that the brain can also operate in a kind of twilight zone between these two states. The most common is sleep inertia: that lingering groggy feeling that can persist long after the alarm has yanked you out of a deep sleep. You are awake but your brain refuses to cooperate; as if part of it is still offline. A more mysterious state is cataplexy, where the loss of muscle tone designed to stop you from acting out dreams unexpectedly switches on during wakefulness and makes you fall over. These phenomena might be due to the mixing of different states of consciousness.

WHY WE CAN'T STAY AWAKE 24/7

FOR MANY YEARS, the question of why we sleep was something of a scientific embarrassment. It is obvious that getting enough sleep is incredibly important to keep us happy and alert, and for our long-term health and well-being, but no one could put their finger on exactly why. In the past few years, though, some important clues have begun to accumulate that help to explain why snuggling down is far from a waste of time.

In the short term, a lack of sleep plays havoc with your emotions and with your ability to make rational decisions. In the longer term, it is implicated in depression, bipolar disorder and schizophrenia, as well as Alzheimer's disease.

Deep cleaning

Deep sleep seems to be particularly important to overall brain health. People with a rare genetic mutation that stops them from entering deep sleep experience dementia and die a couple of years after symptoms first appear. We now know why this might be. A previously unknown waste-disposal system was recently discovered in our brains.

It is made up of cells called glia. These cells act as a kind of personal assistant to neurons, providing energy and nutrients and, importantly, clearing up any waste afterwards. Just as the lymphatic system drains toxins from your body, this 'glymphatic system' collects and breaks down metabolic debris generated by the hard work of the neurons and flushes them out through the cerebrospinal fluid that bathes the brain and spinal cord. The glymphatic system is working most of the time, but during deep sleep it goes into overdrive, expelling ten or twenty times the amount of waste.

The clean-up is remarkably thorough: during deep sleep, the brain's glial cells shrink in size by 60 per cent, creating greater space for the cerebrospinal fluid to jet-wash every nook and cranny. Getting enough sleep, particularly the deep stages, may be crucial to cleaning.

This becomes even more important when you consider that one of the key waste products flushed out during this process is amyloid protein, which forms the plaques in the brain that contribute to Alzheimer's disease. Getting enough sleep could be crucial to washing these proteins away, to prevent them from hanging around where they don't belong.

Sleepy memories

Another important role of sleep is to allow memories to move from short-term storage to a more permanent position, which lets new memories take their place the next day. It has been known for a long time that people remember facts better after sleeping and that a lack of sleep hammers recall, but it wasn't known how sleep manages to help us to organise our memories for future reference.

Animal studies suggest that, with the benefit of being disconnected from the environment, the brain is able to weaken new connections it has made during the day, readying the memory banks for tomorrow. Where they go is still a mystery but there are some ideas. In studies where people's brains were imaged as they learned a sequence of buttons to press, scientists were able to record a pattern of activity that related to the memory of the sequence. Before sleep this trace was recorded in the cortex, the outer wrinkly part of the brain, but during the slow brainwaves of deep non-REM sleep this temporary trace faded to be replaced by a similar pattern in deeper parts of the brain.

Many of the details are yet to be filled in, but it seems that non-REM sleep is all about the brain's housekeeping. REM sleep may have a different purpose, helping us to process emotional experiences without the hormonal rush that accompanied it first time around.

Some scientists also think that REM sleep helps us to re-file memories in relation to things that we already know, turning new lessons into an ever-accumulating wisdom.

Restocking

Other possible functions of sleep include DNA repair, which keeps cells of the body and brain functioning and a chance to restock with essential neurotransmitters. It might also give us a rest from physical activity to save energy, and from the mental taxation of social interaction.

Adequate sleep is what keeps our brains healthy, sharp and ticking over as they should. Miss it at your peril.

WHAT HAPPENS WHEN WE GO WITHOUT?

There have been several attempts to stay awake for as long as possible. The reigning world record-holder is Randy Gardner, a sixteen-year-old high school student from San Diego, who in 1964 managed to stay awake for 11 days and 25 minutes as part of a school science project.

Gardner suffered no long-term effects from his attempt, but his brain went increasingly haywire as the days went by. After a couple of missed sleeps his speech became slurred, and he became increasingly irritable. Then he started to become paranoid and have dream-like hallucinations. By the end he struggled to pay attention to anything. Since then *The Guinness Book of Records* has stopped recording sleep deprivation attempts because of the mounting evidence of risks to health. So, really, don't try this at home.

CAN YOU SLEEP
TOO MUCH?

IF SLEEP IS so important for a properly functioning brain, then it makes sense that we should maximise the brain benefits by trying to get as much as possible. But be careful. Sleeping longer is only a healthy option if you are not getting enough in the first place. If you are, then adding more sleep could actually be bad for your health.

Getting the right amount is tricky, not least because there is no magic number that works for everyone. The idea that we should sleep for eight hours a night is often repeated, but rarely backed up with hard data. It doesn't stem from rigorous scientific research but from surveys that ask people how long they tend to sleep each night.

Most people tend to report seven to nine hours, with eight as a rough average, which might be why it has become a rule of thumb. People often overestimate how long they have slept, however, because it's impossible to look at the clock at the exact time you drop off.

Various lines of evidence point to seven hours a night being a good target. Regularly getting less than seven hours has been linked with poor health, increasing the risk of obesity, heart disease, depression, and early death.

Too much? Too little?

Perhaps more surprisingly, given the well-known benefits of sleep, getting more than eight or nine hours seems to be almost as bad. Many studies have shown that people who regularly sleep longer than eight hours have an increased risk of poor mental and physical health and are at greater risk of an early death.

Why having too much of a good thing is bad for you is a bit of a mystery. It doesn't seem to be because people who sleep more are already unhealthy. In long-term studies where people were assessed as healthy at the beginning, and their sleep tracked for years, those who slept too much became less healthy over time and died earlier, suggesting it was the oversleeping that caused ill health and not the other way around.

Some clues come from the fact that both too-much and too-little sleep are linked to inflammation, part of the immune response that is a predictor of everything from depression to diabetes and heart disease. Exactly why getting too much sleep would make the body react in this way isn't yet clear, but it certainly seems that, like eating and even drinking water, having too much of any good thing is bad for us.

What's more, we might not even enjoy getting that extra time in dreamland. Getting too much sleep has recently been linked to an increased likelihood of nightmares, perhaps because it allows more time for REM sleep, when dreams, including nightmares, occur.

Lazing around

Part of the effect of too much sleep could come down to the simple fact that when we are asleep we are lying down and not moving very much. Our brains and bodies function best when we are on the move and when we lie around both are less efficient.

The long-term effects of bed-rest are not appealing: apathy and depression, muscle wastage, an increased heart rate and sluggish digestion to name a few.

The lazing around doesn't even have to be that extreme to have an effect. Even a couple of hours spent sitting in front of the TV has been shown to have an impact on health. So perhaps too much sleep is bad for us only because we spend less time moving and more tucked up in bed. And the fact that more time asleep equals less time in the day to fit in healthy and brain-boosting exercise.

Getting it right

As well as getting the right amount of sleep, it also seems to matter when you get it. Depriving yourself of sleep in the week and then bingeing at the weekends has been linked to bad moods, fatigue and irritability as well as an increased likelihood of cardiovascular disease, even if there is only an hour's difference between going to bed and waking up between weekdays and the weekend.

The only way to guard against this yo-yo effect of too little or too much sleep is to go to sleep and wake up at the same time every day, whatever day it is and whether or not there is somewhere you need to be.

Keeping regular hours will also makes it much easier to work out whether you are lethargic during the day because you have slept too much or because you haven't had enough. If you've been strict for a few weeks and still feel awful, experiment with adding and losing an hour to see how it makes you feel. There is no one size fits all answer, but the health benefits of a refreshing sleep make it well worth the effort of experimenting.

But one experiment not to try at home is that carried out by space agency NASA, in which volunteers stayed lying down for more than a month (to study the effects of microgravity on the body). Rising out of bed again was a painful experience. One volunteer said 'this is what it must feel like to be about 120 years old'. Sleep junkies, beware.

IS MY SLEEP NORMAL?

It's difficult to say. The amount of sleep we need is influenced by our genes and varies from one person to the next. Exactly which genes are involved is not well understood, but one recent study of over 50,000 people found a gene variant that added 3.1 minutes of sleep for every copy you have.

A tiny minority of people, perhaps less than 3 per cent, have a genetic variation that allows them to get by on four to six hours of sleep with no problems at all. When genetically engineered into mice, the gene seemed to let them whizz through the non-REM stages of sleep faster than non-engineered mice, and to recover more quickly from sleep deprivation. This raises the tantalising possibility that we could one day genetically engineer our way to a shorter night's sleep, without the downsides.

WHEN SLEEP
GOES WRONG

EVER SHOUTED AT your partner while you slept, or woken up unable to move? From sleep eating to exploding heads, there is a dizzying array of sleep disorders. Here are some strange things that go bump in the night.

Exploding head syndrome

This entails the sensation of a loud noise, like an exploding bomb or a gunshot, as you drift off or wake up. It affects about 1 in 10 people and nobody knows what causes it. Despite its name, the condition is harmless.

Hypnagogic jerks

These are jumps or twitches experienced as you nod off, often accompanied by the sensation of falling. The cause remains a mystery, but one idea is that hypnagogic jerks arise if you start dreaming before your body becomes paralysed, or that they are a by-product of your nervous system relaxing as you drift off.

Sleep paralysis

A terrifying experience, where the body, which naturally becomes paralysed during REM sleep, is still paralysed when you wake. You are fully conscious but cannot move or speak, sometimes for several minutes. Some people also feel as if they are choking or their chest is being crushed and they may have visual hallucinations. The condition can be exacerbated by sleep deprivation and certain drugs.

REM sleep disorder

If you've ever punched or shouted at your partner in the night, only to remember nothing the next morning, you may have been in the grip of this condition. Here, the body isn't fully paralysed during REM sleep, so people act out their dreams, perhaps hitting out at their partner. It tends to happen only with bad dreams.

Sleep eating

If you wake up feeling full and find that your fridge has been raided, you could be experiencing this rare sleep disorder. With little or no conscious awareness, people with this condition sleepwalk to the kitchen in the night, and can even cook up and consume a large meal.

Somnambulism

Sleep walking – also known as somnambulism – is the acting out of complex behaviour while sleeping. Actions can be anything from just stumbling about to driving or doing the laundry. It occurs when a person becomes stuck in a limbo between sleep and waking. A sleepwalker has no conscious awareness, and won't remember anything. Stress and alcohol can provoke it, but it mostly comes down to genes. Most sleepwalkers have a family history. It is more common in children than adults.

WHAT MAKES A
GOOD NIGHT'S REST?

IT DOESN'T MATTER how much effort you put into getting to bed at a regular time if you struggle to drop off when you get there. So, what are the best ways to ensure a decent night's sleep? It isn't simply a case of going to bed when it gets dark. Studies of modern hunter-gatherers suggest that they, too, stay up long after sunset and get up early.

What we are doing, and they are not, though, is to mess with our circadian clock by shining light into our eyes far too late into the evening. The circadian clock is set by changing light levels between day and night, which are detected by the eyes and sent to the body's suprachiasmatic nucleus, which orchestrates the sleep–wake cycle.

Under natural conditions, the sleep hormone melatonin would begin to rise in the evening as light levels fall. Sitting around a fire won't affect this because fire emits mostly red light, which has little effect on melatonin production. Mobile phones, tablets and laptops on the other hand generate lots of short wavelength blue light, which interferes with melatonin production by tricking the brain into thinking that it is still daylight.

Using screens for two hours before bed reduces melatonin

concentrations by almost a quarter, which not only means it takes longer to fall asleep but has knock-on effects throughout the night. Screen time before bed cuts down REM sleep, perhaps because messing with melatonin early in the night delays the whole sleep cycle, leaving you less time to get through all the stages before morning. Bright, energy-efficient LED light bulbs also release a lot of blue light – so are best avoided in the hours before sleep.

If sitting in a darkened room getting progressively sleepy doesn't seem like much of an evening, you can always turn on the TV. While the light from the TV screen is bright, we normally watch from far enough away that the intensity of the light is too low to have much of an effect. Or, if you just can't put the tablet down, you could have an app that strips out the sleep-robbing blue light, or try blue-light filtering glasses.

Hot and cold

When you finally shut off the screen and reach the bedroom, the next most important factor is temperature. Melatonin cools the body by a couple of degrees while we sleep, and an over-heated bedroom can interfere with this process.

Too cold a room is also a problem, because it forces the body to expend energy trying to keep warm, at the expense of letting you rest. In an ideal world, the bedroom would mimic the body's natural temperature changes by starting the night pleasantly warm, cooling off a little in the middle of the night, and then warming up first thing to allow you to spring out of bed.

If that's too much for even the smartest thermostat to handle, a general rule is to keep the bedroom between 18 and 21°C, with a window open if it's not too noisy.

Drinking problems

A nightcap might help you drop off but inevitably comes at the expense of good quality sleep later on in the night. Having a few alcoholic drinks before bed disrupts slow-wave sleep, boosting alpha brainwaves that are normally only active in the daytime.

Even one drink earlier on in the evening, say at 5 or 6 p.m., seems to disrupt sleep later in the night. This might be because of the way that alcohol is metabolised, releasing chemicals that act as stimulants when your body and brain should be resting.

A more effective nightcap might be sour cherry juice, a supplement rich in melatonin. It's early days but studies suggest that drinking it regularly bought healthy adults an extra half hour of sleep – enough to tip them over the magic seven hours sleep recommended for good health. Plus it cut down their need to nap during the day.

Melatonin pills have shown mixed results, perhaps because they are broken down in the body very quickly: the half-life of melatonin ranges from thirty minutes to two hours, so may not provide a lasting effect.

Bad habits

The bad news is that all of our bad habits seem to play havoc with a good night's sleep. Drinking coffee in the evening, unsurprisingly, makes it harder to get to sleep and to stay asleep and also has a knock-on effect on melatonin production the next day, perhaps lining you up for a second bad night.

Smoking at any time during the day reduces total time asleep by 1.2 minutes – an effect that animal studies suggest is caused by nicotine's disruptive effects on a circadian clock protein in the lungs and brain. And eating late at night, particularly fatty foods, also makes it more difficult to drop off and affects sleep quality throughout the night.

A good night's rest requires good sleep hygiene. A regular bedtime, not too much bright light in the evening and not too much of what you fancy. Ideally save the bright lights and the caffeine until morning.

ANIMAL SLUMBER

Birds, fish, reptiles and other mammals all sleep. Even fruit flies and nematode worms experience periods of inactivity from which they are less easily roused, suggesting sleep is a requirement of the simplest of animals. But surveying the animal kingdom reveals a bewildering diversity in sleep patterns.

Some bats spend twenty hours a day slumbering, while large grazing mammals tend to sleep for less than four hours a day. Horses, for instance, take naps on their feet for a few minutes at a time, totalling only about three hours daily. Wild elephants average just two hours of sleep a night, the least of all mammals. In some dolphins and whales, newborns and their mothers stay awake for the entire month following birth.

CAN I FILL IN THE GAPS WITH NAPS?

NOT GETTING ENOUGH sleep at night? Is there a way to cheat the system by adding a few well-timed naps? While we don't know whether napping could ever be a healthy long-term replacement for nightly sleep, taking forty winks can certainly improve your mental performance in the short term.

The kinds of benefits that you get depend on how long you nap for, and how deeply you sleep. A 'nano-nap', lasting just ten minutes for example, can boost alertness, concentration and attention and the effects can last for as much as four hours afterwards. This is the kind of power nap you should be aiming for if you need to take a break during a long and boring drive.

If you need to be especially on the ball when you get to your destination, then perhaps shut off for around twenty minutes. A slightly longer nap will increase your powers of memory and recall, but, since you are unlikely to enter the deeper stages of sleep, you are less likely to suffer sleep inertia – that lingering groggy feeling that follows waking from deep sleep. On the flip side, waking up before you hit deep sleep does mean that you won't enjoy its restorative benefits.

Learning in your sleep

Thanks to its memory-consolidation role, deep sleep provides the biggest boost to learning. If that's your aim, you're better off settling down properly for a nap of between sixty and ninety minutes. Research shows this aids learning by shifting memories from short-term storage in the brain's hippocampi to lockdown in the prefrontal cortex – a bit like clearing space on a USB memory stick. As well as helping you to retain factual information, longer naps can increase motor memory, which is useful for training skills such as sport or playing a musical instrument.

A longer nap could also improve your ability to cope with tricky colleagues or loved ones. Sleeping for forty-five minutes or more should take you through a stage of REM sleep, which has a role in emotional processing. Brain scans of people following a REM sleep nap showed more positive responses to images and to pleasant experiences. So, if you're feeling emotional, a longer nap might be in order. Just remember to factor in twenty minutes or so afterwards to get over the worst of the sleep inertia.

REM or not

It's also worth bearing in mind that the time of day you nap may affect the type of sleep you get. During the night, each ninety-minute sleep cycle includes a bout of non-REM sleep followed by REM sleep. However, deep non-REM sleep tends to dominate in the first half of the night, with the balance then shifting to REM sleep. A morning nap is much more likely to contain REM sleep as that is the last state the brain was in so, potentially, could contain more emotionally calming dream sleep. Afternoon naps are more likely to send you into slow-wave deep sleep, which is better for more restorative and memory-boosting purposes. Still, there's no guarantee that trying to hack your sleep to pick the benefits will work, because your brain may just take the kind of sleep it needs.

However you play it, the urge to nap is nearly always strongest after lunch. Contrary to popular belief this isn't the fault of a a full stomach, but your circadian rhythms, which naturally dial down alertness between 2 p.m. and 4 p.m. If you're tempted to nap, the best results come from getting the environment right. Find a warm, dim and quiet place to lie down (getting to sleep when you're sitting takes 50 per cent longer). And if you want to keep it short, simply drink a cup of coffee immediately beforehand – the caffeine kicks in after about twenty minutes, wiping away the sleep inertia and leaving you raring to go.

Extreme napping

Some people take the idea of napping to extremes, ditching full nights in bed in favour of micro-sleeps. There are several such programmes, including the Uberman sleep schedule, which involves napping every four hours for twenty minutes at a time, and totalling just two hours sleep a day. It's one of a number of sleep schedules that promise to maximise our waking hours and cut down sleep to its bare minimum. But the jury is out on whether messing with sleep to such an extreme is a good idea.

Proponents claim that it buys them up to five hours of extra productive time a day and that, after a period of adjustment, they feel fine. Critics, though, point out that a full night's sleep allows our brain to cycle several times through a number of phases, each with their own restorative properties. We might be able to cheat time, but perhaps at the cost of these health benefits. We know that people who regularly don't get enough sleep die younger, so there is always the possibility that the hours you save might have to be paid back at the end of your life.

IS IT NATURAL TO SLEEP IN TWO CHUNKS?

In experiments in the 1990s, people were put into a dark cave all day to see what happened to their sleep patterns. After a few weeks they settled into a pattern of two bouts of four hours, with an hour or so of wakefulness in between. Historical evidence also showed that pre-industrial civilisations tended to have two sleeps with a break in the middle. Could this be our natural sleep pattern?

That idea has been challenged by studies of hunter-gatherer tribes in Africa and South America. These people sleep more like us, staying up after dark and then sleeping in one chunk until morning. The handful of studies comparing the two different sleep patterns found that splitting sleep into two chunks made people sleep more deeply but wake up more often and to feel more sleepy overall.

GETTING INSIDE THE DREAMING MIND

MOVE OVER FREUD. Modern sleep researchers are getting a grip on our dream content, and it has nothing to do with our mothers.

Before the discovery in 1953 of REM sleep and its clear link with dreaming, the study of dreams was almost the exclusive domain of the Freudians, who believed the content of our dreams was a hotline to our innermost desires and feelings. But the obvious mental and physical signatures of REM sleep – rapid eye movements, frantic brain activity, and (to the delight of the Freudians) penile erections – opened the door to studying dreams as a real biological phenomenon.

Probing the deep

It wasn't as easy as sleep researchers had hoped, not least because it's difficult to study a state of mind when not even the person experiencing it can put their finger on what it's like.

Dreams tend to be silent movies – with only half containing traces of sounds. They take many forms, from pure perceptual experiences to simple images or unfolding narratives and intense poetic visions. Most of what we dream about is at least loosely related to what is going on in our lives. But rather than appearing in our dreams as single events on action replay, our memories emerge piecemeal, as small fragments that get added to the storyline of a dream, and not necessarily in chronological order. We remember very little of these night-time mental wanderings.

Filling the gaps

What is going on in the brain to generate these experiences? A combination of brain-imaging experiments and, more simply, waking people up to ask what they were dreaming is beginning to fill some of the gaps.

A ground-breaking 2017 study revealed that we dream for a staggering 95 per cent of REM sleep, and, more surprisingly, that dreams occur in much of our non-REM sleep too. There's a big difference in the dream quality from these two types of sleep: REM dreams tend to be more vivid and memorable, whereas the non-REM variety are shorter and duller.

During dreams, the brain's visual areas are very active, as are the amygdala, thalamus and the brainstem, which fits with the fact that dreams tend to be vivid and emotional.

One big question is whether the brain activity you see in sleep corresponds with the experiences we call dreams. It turns out it does. When your dream involves people's faces, for example, the part of your brain's visual processing system that deals with facial recognition shows intense activity, and a dream with speech in it lights up your brain's Wernicke's area, which is responsible for the comprehension of speech. What's more, being able to later remember a dream is linked to higher activity in the prefrontal cortex – which is associated with memory – while dreaming.

Neuroscientists have even found the characteristic brainwave pattern of the dreaming brain. During sleep, low-frequency brainwaves are detectable across the brain. A decrease in these waves in a particular area at the back of the brain is a sign that someone is dreaming.

But why?

None of this, however, answers the big question of why we dream. One common explanation is that it helps to forge links between the events of the day and what is already in the memory, allowing us to make sense of aspects of them in the wider context of our lives. As part of this process, our brains might dredge up old memories and plant them in our dreams, which might be why we often dream of people and places that we haven't seen for years. But no one knows whether dreaming is essential for preservation of our memories – or could we manage to store our life's events without them?

Dreams seem to play a central role in our emotional lives too. Emotions affect not only the flavour of the dream but also how likely it is to stick on waking – and they may also help us to come to terms with difficult events. Sleep, and REM sleep in particular, selectively strengthens negative emotional memories. This sounds like a bad thing – but the upside of remembering bad experiences is that you can learn from them. Evidence for this comes from a series of studies starting in the 1960s, following people who had gone through divorces, separations and bereavements. Those who dreamed most about these events later coped better, suggesting that their dreams had helped.

In addition, reliving the upsetting experience in the absence of the hormonal rush that accompanied the actual event helps to strip the emotion from the memory, acting as a kind of balm for the brain. In people with post-traumatic stress disorder this emotion-stripping process seems to fail for some reason, so that traumatic memories are recalled in all their emotional detail – with crippling psychological results.

But it is also possible that dreams don't actually mean anything at all. They could just be an epiphenomenon, or side effect, of brain activity going on during sleep. It is this underlying neuronal activity, rather than the actual dreams, that is important. Don't tell Freud.

SCIENTIFIC DISCOVERY WITH YOUR EYES SHUT

'What if in your sleep you dreamed,' asked the poet Samuel Taylor Coleridge, 'and what if in your dream you went to heaven and there plucked a strange and beautiful flower, and what if when you awoke you had the flower in your hand?'

There have been many reports of people waking from dreams in possession of a 'flower', in the sense of an artistic creation or a solution to a problem, which they 'plucked' in their dreams. The best known is the experience of the German chemist Friedrich August Kekulé, who for many years had been trying to determine the molecular structure of benzene. One night in the winter of 1861/2 as he dozed in front of a fire, he dreamed of snake-like chains moving and twisting. Then one of the snakes grabbed hold of its own tail. He awoke and realised that the structure of benzene was a closed carbon ring.

CONTROL YOUR DREAMS

IMAGINE BEING ABLE to shape your dreams so that you can fly or visit the past. It might sound like mumbo jumbo but the idea that we can knowingly experience and control dreams has been scientifically verified.

If controlling your dreams sounds intriguing, here are some tips on how to make it more likely to happen to you.

Reality test

You might improve your chances of lucid dreaming by training yourself to conduct frequent reality tests while you are awake, such as counting the fingers on your hand, reading and re-reading the words on a page or turning the lights off and on again.

Later you might find yourself doing the same while you are dreaming. If you do, you're more likely to notice when, in a dreamworld, your hand looks unusual, words jump around on the page or light switches don't work.

Plan your dream

Almost as fun as the dreaming itself. Before you go to bed, think about what you want to dream lucidly about, in as much detail as possible.

There is evidence that we are more likely to incorporate into our dreams some direct types of stimulation – such as pressure on our limbs, the spray of water and the uttering of meaningful words. The tricky bit is that you need to be at the REM stage of sleep for this to work. You could ask a friend to stay up and wait for your eyes to flicker.

Hit snooze

One survey of lucid dreamers found that there was a correlation between hitting snooze on the alarm clock and experiencing a lucid dream. The catch is that the usual ten minutes between pressing snooze and the next alarm isn't really long enough to send you back to dream sleep.

To get over this problem, you could set two separate alarms, one later than the other. Leave at least half an hour, preferably more, between them to give yourself a decent chance to begin dreaming again.

Total recall

When you wake up, try to recall as many of your dreams as you can, to practise crossing over between the dreamworld and reality. Good luck – and sweet dreams.

13.

TROUBLESHOOTING

THE PITFALLS OF
SLOPPY THINKING

FAKE NEWS, CONSPIRACY theories, folklore — sometimes it seems that we are less inclined to think sensibly than ever before. Given we pride ourselves on being *Homo sapiens*, the thinking ape, how did it come to this? The truth is, our ancestors evolved a whole host of cognitive shortcuts that helped them survive. The problem is that our world is very different and as a result the ways of thinking that come most effortlessly can lead us down the garden path. To avoid these pitfalls, you have to identify them. So here's a guide to sloppy thinking.

Folk knowledge

Children, it is often said, are like little scientists. What looks like play is actually experimentation. They formulate hypotheses, test them, analyse the results and revise their world view accordingly. That may be true, but if kids are like scientists, they are rubbish ones. By the time they enter school, they have filled their heads with utter nonsense about how the world works. The job of education is to unlearn these 'folk theories' and replace them with evidence-based ones. For most people, it doesn't work. No wonder the world is so full of nonsense.

In biology, for example, young children often conflate life

with movement, seeing the sun and wind as alive, but trees and mushrooms as not. They also see purpose everywhere: birds are 'for' flying and rain falls so flowers can drink. In physics, children conclude that heat is a substance that flows from one place to another, that the sun moves across the sky, and so on. For everyday purposes, these ideas are serviceable. Nevertheless, they aren't true. Children cling to their folk theories, and when they encounter difficult concepts, they cling even harder. For example, many intuitively see evolution as a purposeful force that strives to endow animals and plants with the traits they need to survive.

Researchers have shown that folk theories can be suppressed by a more scientific world view, but cannot be eradicated altogether. In one study, people were presented with a variety of statements about the natural world and asked to say which were true and which false. Some were designed to be intuitively true but scientifically false, such as 'fire is composed of matter'; others were intuitively false but scientifically true, such as 'air is composed of matter'. People who got the right answer still took significantly longer to process an intuitively false but scientifically true statement. This was even the case for those who had been scientists for decades.

Similar results come from brain scans. When people watch videos that are consistent with the laws of physics but intuitively wrong – such as light and heavy objects falling at the same rate – the error-detecting parts of their brains light up, suggesting that they are struggling to reconcile two competing beliefs. The upshot is that scientific thinking is hard-won and easily lost, and that persuading most people of the validity of things like evolution, climate change and vaccination will always be an uphill struggle.

What are you looking at?

We are born to judge others by how they look: our brains come hardwired with a specific face-processing area, and even shortly after birth, babies would rather look at a human face than anything else. By the time we reach adulthood, we are snap-judgement specialists, jumping to conclusions about a person's character and status after seeing their face for just a 10th of a second. And we shun considered assessments of others in favour of simple shortcuts – for example, we judge a baby-faced individual as more trustworthy, and associate a chiselled jaw with dominance.

Unfair, it may be, but it makes good evolutionary sense. Being able to quickly assess whether someone is friend or foe is important survival information. But there is a problem. More often than not, our first impressions are wrong. It's not clear why, the fact that we meet many more strangers than our prehistoric ancestors probably plays a part.

Another problem is that we don't stick to stereotyping faces one at a time. Studies show that we are just as quick to categorise groups of people – and then discriminate against them as a result. These findings don't paint us in a great light. We tend to dehumanise groups we judge to be lacking in warmth, and react against high status rivals, sometimes violently, based on feelings of envy (historically, many genocides fall into this category). And there are more downsides; for instance we may pity those of low status, but react by patronising them, and the pride we feel towards our own group can spill over into nepotism.

If you think you are above this kind of thing, think again. Even if you consciously reject stereotypes, the culture you live in does not, and experiments suggest that you are likely to share its biases. One study, for example, found that white Americans who showed no sign of racism on a standard test subconsciously dehumanise black people. The best way to escape this evolutionary trap is to get to know people from outside your echo chamber. Working together on a joint project is ideal because relying on someone forces you to look beyond simplistic first impressions.

Suckers for a celeb

If you ever meet the Queen, there are certain rules you are advised to follow. Do not speak until spoken to. Bow your head, or curtsy. Address her first as 'Your majesty', then 'Ma'am', but 'Your majesty' again upon leaving. Don't make the mistake of calling her 'Your royal highness' – that is for other members of the royal family.

Apply some rational thought and this is all very puzzling. What has the Queen done to deserve such treatment? If humans were a species of primate, you would conclude that the Queen must be the dominant female. But dominance has to be earned and kept, often by physical aggression. These days nobody defers to the Queen out of fear that she will beat them up if they don't. Human societies do have dominant individuals, but what the Queen possesses is something quite different: prestige. And we are suckers for it. According to biologists, this prestige bias is an evolved feature of human cognition that goes back to the time when our ancestors were nomads living in small bands. Humans are social learners, which means we copy the behaviour of other people. People who copy successful individuals can acquire useful, survival-enhancing skills – how to hunt, for example. But to do so requires close contact with the skilled, without getting on their nerves. The best way to do this is to 'kiss up'. Pay them compliments, do them favours, sing their virtues and exempt them from certain social obligations. Evolution thus favoured sycophants.

Prestige exerts such a strong pull on the human mind that the construction and perpetuation of hierarchies is hard to resist. In lab experiments, people find it easier to understand social situations where there is a clear pecking order, and they express preferences for hierarchies, even if they are at the wrong end of them.

Strong beliefs

If God designed the human brain, he (or she) did a lousy job. Dogged by glitches, requiring routine shutdown for maintenance for eight hours a day, and highly susceptible to serious malfunction, a product recall would seem to be in order. But in one respect at least, God played a blinder: our brains are almost perfectly designed to believe in him/her. Almost everybody who has ever lived has believed in some kind of deity. Even in today's enlightened times, atheism remains a minority pursuit requiring hard intellectual graft. Even committed atheists easily fall prey to supernatural ideas. Religious belief, in contrast, appears to be intuitive.

Cognitive scientists talk about us being born with a 'god-shaped hole' in our heads. As a result, when children encounter religious claims, they instinctively find them plausible and attractive, and the hole is rapidly filled by the details of whatever religious culture they happen to be born into. When told that there is an invisible entity that watches over them, intervenes in their lives and passes moral judgement on them, most unthinkingly accept it. This is not brainwashing. Many scientists argue that religious belief is a side effect of cognitive skills that evolved for other reasons. It pays, for example, to assume that all events are caused by agents. The rustle in the dark could be the wind, but it could also be a predator. Running away from the wind has no existential consequences, but not running away from a predator does. Humans who ran lived to pass on their genes; those who did not became dinner.

Theory of mind

Then there's theory of mind, which evolved so that we could infer the mental states and intentions of others, even when they aren't physically present. This is very useful for group living. However, it makes the idea of invisible entities with minds capable of seeing into yours quite plausible. Religion also piggybacks on feelings of existential insecurity. Randomness, loss of control and knowledge of death are soothed by the idea that somebody is watching over you.

It has even been argued that religion was the key to civilisation because it was the social glue that held large groups of strangers together as societies expanded. But these days religion's downsides are more apparent. Conflict, misogyny, prejudice and terrorism all happen in the name of religion. However, as the rise of atheism attests, it is possible to override our deep-seated religious tendencies with rational deliberation – it just takes some mental effort.

Wanna be in my gang?

A tribal mentality is frighteningly easy to induce. Researchers took twenty-two adolescent boys to what seemed a traditional summer camp, but in truth it was a psychology experiment. The boys bonded as two groups, each unaware of the other's existence. Then the experimenters engineered a fleeting encounter between the groups. Despite the boys having been chosen for their similarities, the camp descended into a tribal warfare, with insults, land grabs, even a mass brawl. Hostilities ended when a common enemy was introduced in the form of fictitious vandals.

Tribalism can be a motivating force: rivalry between scientific teams working on the same problem, for instance. But it also underpins unedifying behaviours including racism and homophobia. Our saving grace is that our tribal boundaries are fluid. Fans of rival football clubs align to support a national team, for example. If we can extend our definition of the tribe in football to include other groups, why not in other, more meaningful, areas of life?

GETTING YOUR OWN BACK

It is, according to popular wisdom, a dish best served cold. Revenge appears to be a universal human trait. A study of ten hunter-gatherer groups found that all of them had a culture of vengeance – with a common list of wrongs that needed avenging.

The desire to inflict punishment makes sense. The original wrong cannot be righted, but the revenge is a social signal that makes others think twice about following suit. But undercook revenge and you reveal that you are worth exploiting. Overcook it and you risk starting a tit-for-tat cycle of revenge, which is in nobody's interest. The fact that we often make such misjudgements might explain why we have evolved an instinct for forgiveness, too – to minimise the fallout.

In modern societies, revenge is normally delegated to the state. Still, many people take it into their own hands. Revenge is a major motivation for terrorists. And it is a causal factor in up to 20 per cent of homicides worldwide. All of which suggests that revenge might in fact be a dish best avoided.

THE MIND-SLIPS THAT LEAD TO CATASTROPHE

YOUR BRAIN IS capable of great creative feats – but also the odd catastrophic piece of decision-making – which in our complex world can lead to major disasters. Fortunately, our growing understanding of what makes us tick is giving us new ways to avoid these glitches – and so harness our minds to avoid damage to life and limb.

Confirmation bias

When BP's Deepwater Horizon oil drilling rig exploded in 2010, the flames were visible thirty miles away. Before the blowout, staff had tested the concrete seal on a freshly excavated well. The results indicated that the seal was not secure and removing the underlying column might result in a catastrophic blowout. So why were the signs ignored? Disaster analysts later found that the workers viewed the test as a means of confirming that the well was sealed, not finding out whether it was or not. When the test failed, workers explained it away. This reluctance to face facts is nothing unusual. Most of us have trouble believing evidence that contradicts our preconceptions. Psychologists call this confirmation bias.

Dopamine – our brain's reward hormone – may be to blame.

Acting on the front of the brain, it makes us inclined to ignore evidence that challenges long-held views, keeping us from having to revise constantly the mental shorthand we use to understand the world. In another part of the brain, the striatum, dopamine has the opposite effect: its level spikes in response to novel information, and that makes us more likely to be open to these details. The net result of the two effects is to favour established beliefs. If you want to cut out confirmation bias from your decisions, think about putting across a counter-argument, forcing yourself to consider alternative points of view.

Fixation error

In 2005, Elaine Bromiley went into hospital for a sinus operation. When her airway became blocked, three doctors tried to insert a tube down her throat. When that failed, they should have performed a tracheotomy, cutting open her windpipe so that she could breathe. Instead, the doctors kept trying to get the tube in, not noticing that their patient was being starved of oxygen. She never woke up again.

This type of mistake – fixation error – occurs because we have a remarkable ability to focus attention on the things we care about or that are relevant to our task. But sometimes it means we miss things. The aviation industry has fought this by encouraging communication among crew. If one person misses something, others can point it out. Before this culture was introduced, crew sometimes felt unable to challenge a captain about a problem. Operating theatres in hospitals have followed suit. During Elaine Bromiley's surgery several nurses noticed that she was turning blue but felt they couldn't tell the doctors what to do. Now UK public hospitals have implemented checklists in an attempt to reduce bad communication and prevent similar disasters.

Primal freeze

Fear evolved as a survival mechanism. When we encounter danger, our hearts race and the stress hormone cortisol floods our system, giving muscles access to extra energy. The trouble is that cortisol also knocks out working memory, which allows us to process information and make decisions, and declarative memory – our ability to recall facts and events. From an evolutionary standpoint, this makes sense. When you're running from a tiger, it's not important to remember how you did it. But in our complex modern world, where cognitive dexterity can be more important than physical feats, our fear response can leave us compromised. That doesn't mean that we can't perform under stress. Cortisol doesn't disable procedural memory, which allows us to do things like walk or open a door. Procedural memory is also what allows highly trained pilots to perform under difficult conditions.

But without training, our cortisol-compromised mind may cause us to freeze or engage in behaviours unsuited to the situation. How do we fight the freeze? Practice. Surprises come in all shapes and sizes – the more prepared we are to face them, the easier we can deal with the unexpected.

A wandering mind

Every driver has been there – you hit a quiet stretch of road and your thoughts turn to dinner or an upcoming holiday. As soon as the environment becomes predictable, your mind starts to wander. Daydreaming has been implicated in train derailments and as many as half of all car crashes. When our thoughts drift, brain structures known as the default mode network kick into gear. It seems to play an important role in helping us to organise our thoughts and plan our future. However, that's not necessarily useful while you are operating heavy machinery. Thankfully, there are some strategies to keep your mind on the task.

One is to be aware of your body clock. Early risers pay attention for longer earlier in the day, whereas night owls are better at staying focused in the evening. Drivers may find that taking an unfamiliar route improves focus. Chewing gum and consuming caffeine, too, have been shown to help people stay focused on tedious tasks.

BE A DEVIL

Whether you're aware of it or not, we're all inclined to bend our opinions towards those of the majority. This 'group-think' is thanks to the ventromedial prefrontal cortex, a part of the brain's reward centre that lights up when we encounter things we want, like a chocolate bar. It's also activated when people are told what others think.

Conformity can be useful, letting others serve as a guide in unfamiliar situations. But it can also lead us into danger. In 2012, three members of a skiing group, which included pros and sports reporters, died in an avalanche. A photographer on the trip said he'd had doubts about the outing but dismissed them. There's no way this entire group can make a decision that isn't smart, he'd thought.

How do we avoid such errors? Be the devil's advocate and find ways to spark debate.

YOUR BRAIN ON DRUGS

FOR MUCH OF our existence, humans have used a variety of mind-altering substances to experience the world in different ways. All drugs have an effect on the brain's reward system, giving a sense of pleasure by boosting levels of the feel-good chemical dopamine. But this system can be corrupted into addiction.

Nicotine

This potent and addictive stimulant is found in the dried leaves of the tobacco plant. It boosts attention and memory by activating many brain areas and networks, including the parietal lobe at the top of the brain – important for processing sensory information, knowledge of numbers and the manipulation of objects. Nicotine mimics acetylcholine, a key neurotransmitter for arousal, attention and memory.

LSD

The trippy effects of LSD include hallucinations, feeling separate from the body, and feelings of bliss. The drug causes the brain's visual cortex to connect more strongly to other regions (hence the hallucinations) as well as affecting many other brain areas. It works by latching on to the brain's serotonin receptors.

Marijuana

Paranoia, munchies, giggles, sleepiness. You get these effects from marijuana because of its active ingredient, THC, which hits the brain's natural cannabinoid receptors. These influence pain, appetite, memory and mood. Among other effects, THC alters activity in the brain's memory centre, the hippocampus, causing short- and long-term memory loss.

Alcohol

Unlike many drugs, alcohol affects the majority of the brain's signalling pathways. One brain area on its hit list is the cerebellum, which controls voluntary movements and balance – hence the drunken stagger after a few drinks. Another way alcohol makes you sluggish is by heightening the brain's sensitivity to GABA neurotransmitters, chemicals that dampen neural activity. Paradoxically alcohol also makes you feel euphoric, due to the release of noradrenaline, the chemical responsible for arousal.

Opiates

Drugs such as morphine and heroin are derived from the opium poppy. They block pain signals from the body to the brain by locking onto the naturally occurring opioid receptors, which are widely distributed in the brain and spinal cord.

Caffeine

The world's most popular psychoactive drug, found in coffee and cocoa beans, boosts concentration and speeds up reaction times by activating the brain's prefrontal cortex. It blocks the action of the drowsiness-causing neurotransmitter adenosine, and stimulates the release of the fight-or-flight hormone adrenalin – hence the jitters you feel after a few strong coffees.

Cocaine

The powerful stimulant cocaine gives a high by flooding the brain with the feel-good chemical dopamine. Longer-term use alters the wiring in the frontal area of the brain responsible for decision-making and inhibition, making the drug harder to resist in future.

WHY FALLING IN LOVE
DERAILS THE MIND

ROSES ARE RED. Violets are blue. Do you love me? *Because I'm crazy for you.* There may be more to this than meets the eye. When it comes to love, many psychiatrists believe that passion's thrills resemble mental illnesses – both in their outward habits and the brain's inner chemistry. And their conclusions might just explain why love makes you do such foolish things.

This idea was first hit upon by researchers investigating the cause of obsessive–compulsive disorder (OCD). One of their chief suspects was the chemical messenger called serotonin, which has a soothing effect on the brain. They found that serotonin levels were usually low in people with OCD. But while interviewing these people, they were struck by the way their persistent one-track thoughts mirrored the musings of people in love. Both groups spend hours fixating on certain objects or that certain someone. Both knew their obsessions were irrational, yet couldn't snap out of them. Was serotonin to blame for love as well?

To find out, researchers took blood from people who had fallen in love within the past six months and who had obsessed about their new love for at least four hours every day, but who hadn't yet had sex. They then compared it with that taken from a group of people with OCD, and a group who had neither affliction.

Both the OCD group and the group in love had 40 per cent less serotonin in their blood. When they retested those in love a year later, their serotonin levels had bounced back to normal, while their obsessive giddiness had been replaced with a more subtle affection.

It's not just OCD that love resembles, but addiction, too. In the early stages of a romantic relationship, the brain's reward centres are flooded with dopamine. This gives a high similar to an addictive drug, creating powerful links in our minds between pleasure and the object of our affection, and meaning that we crave the hit of our beloved again and again.

And it's true what they say about love being blind – during the first stages of love, brain regions linked with negative emotions and critical social judgement switch off. Funnily enough, the brain regions associated with emotion, and hormones associated with bonding, only have a role in later phases of a relationship.

Crazy in love

So why this initial craziness? By reeling you in with intense cravings at the beginning of a relationship, making you believe that you've caught the one heart-stopping fish in the sea, the brain keeps love's fires burning long enough for romance to yield an evolutionarily satisfactory end: offspring. Unfortunately, it's not all wine and roses when it comes to love. Falling in love may have evolved to improve our chances of reproduction, but this also means people are predisposed to terrible suffering when jilted by their beloved.

Painful emotions develop when the reward centres of the brain, associated with the dopamine high of falling in love, fail to get their hit. Paradoxically, when we get dumped we tend to love back even harder, as the brain networks and chemicals associated with love increase. Panic also kicks in as we feel something akin to the separation anxiety experienced by young mammals abandoned by their mothers. Then love can turn to anger and hate, as the regions associated with reward are closely linked to rage in the brain. Finally, when jilted lovers are resigned to their fate, they often enter into prolonged periods of depression and despair. These negative emotions can spawn anything from obsession and domestic violence to stalking and even murder of supposed loved ones.

A cure for love

So is there any way to fight the unwanted side effects of unrequited love? Some argue that anti-love solutions could help people struggling with suicidal or delusional thoughts or those in the clutches of unrelenting grief. Drugs that boost serotonin can offer relief to people with OCD, so it's reasonable to think that they could also help to dampen lustful feelings. These drugs include antidepressants called selective serotonin re-uptake inhibitors (SSRIs), which are known to blunt extreme emotions and make it harder to form romantic bonds. This is an unwanted side effect for people with depression, but for those seeking to detach from someone, it could be welcome.

What if it's not lust but the heartache from a lasting bond you want severed? Blocking corticotropin-releasing factor (CRF), a hormone involved in the stress response, stops the depressive behaviour that prairie voles – famously monogamous creatures who form one life-long bond – exhibit when their partner dies.

We probably wouldn't want to block this hormone for unrequited love, but it could be helpful to relieve the depression that comes with persistent grief. In the end, it seems that all aspects of love come down to the right chemistry – inside and out.

HOW TO MEND A BROKEN HEART

Since love shares some of the same neural underpinnings as addiction, fixing a broken heart may involve replacing your fix of oxytocin or dopamine. That doesn't mean you have to pop a pill. Try exercising, increasing your bodily contact with someone new, and going out more with others – all of these things raise your oxytocin levels.

Ultimately, you've got to just give it time. When researchers studied the brain as love fades away, they found that people who are pining after a lost love have greater brain activity in the ventral pallidum, which is involved in attachment, than people who were happily in love. This activity diminished over time, as their attachment also waned. One day it might be possible to use brain stimulation to decrease activity in the ventral pallidum, and speed up these healing effects.

SIX WAYS YOUR BRAIN CAN MAKE YOU FEEL STUPID

EVER SEEN A man in the moon, accidentally called your boss 'Mum' or burst out laughing at bad news? It turns out that brain slips are as common as physical trips. At *New Scientist* we call them brain farts, and here are our favourites.

The doorway effect

We have all walked into a room and immediately forgotten why we'd gone there in the first place. To investigate this common occurrence researchers asked people to navigate a virtual environment. Occasionally the participants would pick up an object, causing it to disappear from view. Now and again they would be asked what they were carrying. If they had moved into a different room, they were slower and less accurate at remembering what the object was. The same thing happened when the experiment was repeated in real rooms.

As we move around the world, our brain constructs a temporary 'event model' of our environment and our thoughts and actions within it. But storing several event models at once is inefficient. New environments may require new sets of skills, and so it is best to focus our memory on what is currently at hand.

Doorways seem to trigger the replacement of one event model with another, which makes us more likely to forget what happened in the first room. It's not just doorways that trigger this shift – passing from rural fields into a town can do it too, or from highways to backstreets, upstairs to down.

Coffee. Coffee. Coffee.

Read a word enough times and not only does the spelling seem impossible but the word starts to lose meaning. This peculiar feeling is called semantic satiation. It's a form of mental flatulence thought to be a result of cellular fatigue. When a brain cell fires, it uses energy. It can usually fire a second time immediately, but if it keeps on firing it eventually tires and must take a short break.

When we read a word over and over, the brain cells responsible for processing all aspects of it – its form, meaning and associations – tire. And so the word stops making sense. More meaningful words, such as 'massacre', may take longer to appear alien because your brain cycles through several associations that it attaches to the word before fatiguing. A less evocative word such as 'coffee' may take only a few repetitions to turn into gobbledygook.

Is that door handle smiling at me?

Kate Middleton recently turned up in a jelly bean, and Jesus in a jar of Marmite.

Seeing faces in inanimate objects is called pareidolia. You've probably experienced it yourself, in the form of the man in the moon. It has a simple explanation: evolutionarily, it makes sense for the brain to be on high alert for faces. We need to be able to detect one and understand its motives in order to react appropriately. That we are occasionally too good at it, spotting the Virgin Mary in grilled cheese, is of little consequence compared with failing to spot a face hidden in the woods.

I mean goxi furl

In 2012, CBS news anchor Robert Morrison referred to Prince William as the 'douche of Cambridge' rather than the 'duke'. Freud would have said that Morrison's slip gave away his thoughts, but there could be a more forgiving explanation. When we speak, the brain calls up numerous networks; those that consider possible word choices, that process meaning, that help us form individual sounds, for instance. With all this processing going on, the brain occasionally makes a mistake, failing to suppress an alternative choice of word or activating the sounds for one word instead of another. Sometimes an entirely inappropriate word pops out, (ever called your boss 'Mum'?). This happens because the word shares some context with the one you intended – your boss might look like your mother, for instance.

That said, there is some evidence to support Freud's ideas. In one experiment, groups of heterosexual men read pairs of words silently until a buzzer sounded, at which point they spoke the words aloud. One group had been greeted by a middle-aged man. Another group was met by a provocatively dressed young woman.

The men all made the same number of slips, but in different ways. The men greeted by the woman made more sex-based slips, reading 'goxi furl' as 'foxy girl', for instance. So on occasion our thoughts do seem to influence our linguistic stumbles. Maybe there was more to Morrison's slip-up after all.

Don't laugh

As faux pas go, it's pretty awkward. In the middle of a row, or when someone tells you some terrible news, the only thing you can do is laugh. One possible reason is that laughter acts as a social glue; it tells the people you are with that you like them and think like them. So an urge to laugh in the middle of an argument may just be an innate way of defusing the situation.

Another theory suggests that the sound of laughter evolved to inform those who share our genes that a situation is a false alarm. Nervous laughter may therefore be a way of convincing ourselves and others that a situation is not as bad as we might think and therefore guarding against any debilitating anxiety that might result from the experience.

So you're not a black-hearted buffoon. Just blame your over-protective brain.

IS THAT REALLY WHAT I SOUND LIKE?

When we speak we hear our voice in two ways. The first is just as others hear us – via sound waves that make eardrums vibrate. The other is via vibrations from our vocal cords that travel through the skull to our eardrums. Both sets of vibrations are transferred into nerve signals that are combined and then processed by the brain to give you an impression of what your voice sounds like.

However, as the vibrations from your vocal cords travel through your skull, they spread out, which lowers their frequency and leads to the impression that they are lower in pitch. When you hear your own voice on a recording, you hear its true pitch – which isn't the sound you have spent your whole life hearing.

WHY WE ARE SO ANXIOUS – AND HOW TO GET AROUND IT

DRY MOUTH? RACING HEART? Knotted stomach? You've got the hallmarks of anxiety. It's a natural response that evolved over millions of years to make us more vigilant and good at fleeing from danger.

Unfortunately, sometimes this high state of alert won't switch off. Anxiety disorders – including panic attacks, social anxiety and phobias – are now the most prevalent mental health problem in Europe and the United States. On average one in six of us will contend with an anxiety disorder at some stage in our lives – women more than men. So what's causing some of us to be natural-born neurotics? And can we do anything to tackle it?

What's normal?

Feeling anxious because you heard a noise on a dark street isn't the same thing as having an anxiety disorder. The key difference is whether anxiety is interfering with your day-to-day life, or causing you distress.

In social anxiety disorder, the most common anxiety disorder, you might believe that blushing will result in people laughing at you. If you have panic disorder, you might assume that you are having a heart attack if your heart starts to race. Generalised anxiety disorder is characterised by chronic worrying about a range of different events, for at least six months.

Anxiety was first recognised as an illness by the ancient Greeks, and has persisted into modern times. In the United States in 1980, the American Psychological Association estimated that between 2 and 4 per cent of people had an anxiety disorder. Today, some studies suggest it's more like 18 per cent in the United States and 14 per cent in Europe. Such figures suggest we are in the midst of an anxiety epidemic, fuelled by factors such as economic insecurity, the demands of social media and the rise of the twenty-four-hour society. The reality is more complex; the apparent increase is probably due to changes in diagnosing the condition over the years, which make long-term comparisons difficult.

Even if the prevalence of anxiety disorders hasn't increased, anecdotal evidence suggests that the type of anxiety people are experiencing is changing. Twenty years ago, the majority of queries received by the charity Anxiety UK were from people with panic disorder or agoraphobia, an extreme fear of open spaces. Nowadays it is hypochondria and social anxiety.

Fight or flight

Recent studies offer some insights into why anxiety seems to take over in some people. Central to it is the amygdala, a brain region that processes our emotions and triggers the release of the hormones responsible for the fight-or-flight response. The amygdala is linked to parts of the brain that process social information and help us to make decisions. During bouts of everyday anxiety, this brain circuit switches on and then off again – but in people with anxiety disorders it seems to get stuck in the on position. Fear memories stored in the amygdala prime us to respond to threats we have previously experienced. This response is normally kept in check by a parallel circuit stemming from the prefrontal cortex, which can temper our learned response. Occasionally the system fails, however. War veterans with post-traumatic stress disorder (PTSD), for instance, have abnormally low levels of activity in their prefrontal cortex, and unusually high levels in their amygdala.

Anxiety prone

Do you calmly navigate life's bumps or agonise at every turn? Psychologists argue that we have innate dispositions that explain how we act, one of which is neuroticism – or proneness to anxiety. A study of more than 106,000 people identified nine regions of the genome that seem to correlate with neuroticism. Some people are therefore naturally more prone to anxiety. Your age and sex may also play a role. Most anxiety disorders peak in eighteen to thirty-four-year-olds before dropping off. And women are about twice as likely to develop an anxiety disorder as men. In part, this may be down to hormones and their influence on the brain.

Tackling the problem

If you have an anxiety disorder, cognitive behavioural therapy (CBT) is likely to be the first recommended treatment. It aims to address the maladaptive beliefs that drive your anxiety. Once they have been identified, CBT helps you challenge them. Therapy isn't for everyone, however. Some people don't respond well to therapists or analysing their own behaviour. In this case, drugs may be able to redress chemical imbalances in the brain.

Exercise can help with day-to-day anxiety and is a helpful strategy to add to your toolkit. It triggers the release of mood-boosting endorphins, and forces you to concentrate on something other than your own thoughts. Then there's diet. A recent study showed that taking a fibre-rich supplement to encourage the growth of beneficial gut bacteria caused people to pay more attention to positive words on a computer screen and less attention to negative ones. Upon waking each morning, the volunteers also had lower levels of the stress hormone cortisol in their blood.

Modern life may be packed with events outside your control, seemingly designed to foster anxiety and self-doubt in your anxiety-prone brain. The important thing is to recognise the symptoms and do something about them.

ALCOHOL AND SOCIAL ANGST

Many of us have a drink to help us feel more relaxed in social situations. Could this have been the reason why alcohol was so highly prized in ancient civilisations? This is the argument put forward by psychiatrist Jeffrey Kahn, who thinks that alcohol may well have been the first widely used psychopharmacological medication – given its ability to 'lubricate' a socially anxious person. It is no secret that alcohol can allow us to disregard our varying degrees of social angst and pursue endeavours that we may otherwise avoid. Kahn thinks this might have an evolutionary purpose too: for a person who is normally withdrawn, 'beer muscles' can provide a chance to get socially involved or contribute to their community in a way they might not otherwise.

14.

UNLOCKING YOUR POTENTIAL

DEFEATING DEMENTIA AND OLD AGE WITH BRAINPOWER

RICHARD WETHERILL WAS a superb chess player. The retired university lecturer could think eight moves ahead – but then his razor-sharp mind started to dull. When he found he could no longer think five moves ahead, he was sure something was wrong and arranged for some neurological tests.

He sailed through every test designed to spot early dementia. Two years later, Wetherill died suddenly, and an autopsy revealed a brain riddled with plaques and tangles – the hallmark of Alzheimer's disease. Most people with that level of damage would have been reduced to a state of total confusion. Yet for Wetherill the only impact was that he could no longer play chess to high standards. What was he doing differently? What was cushioning the blow?

Wetherill's experience is a perfect example of a phenomenon that has puzzled scientists: people who lead more intellectually stimulating lives, who are more intelligent, better-educated and have high-status occupations, are somehow protected from the mental decline that comes with age. This mental padding has been dubbed 'cognitive reserve'. The higher your reserve, the more damage you can sustain without showing signs of mental decline.

Downhill slope?

Dementia isn't inevitable. The human brain can stay sharp well past a hundred years of life. Yes, getting older slows us down: parts of the brain associated with executive function (the high-level cognitive processes that allow us to develop detailed plans and execute them) and memory shrink, myelin sheaths around our neurons start to erode, slowing down signalling, and arteries narrow, diminishing blood supply. But those things mainly affect speed: when healthy older people are given extra time to perform cognitive tasks, the results are on par with the young. In contrast, dementia alters the cognitive playing field. As well as affecting memory, it causes issues with understanding or expressing oneself in language, problems with sensory perception, and disturbances in executive function that can undermine day-to-day independence.

The number of people affected by dementia may be rising but that's largely because more of us are living longer. Between the late 1980s and 2011, the proportion of people over sixty-five with dementia actually dropped by 20 per cent in England and Wales. Between 2000 and 2012, dementia rates in that age group dropped by 24 per cent in the United States. Similar declines have been reported in other developed countries.

Compensation

A boost in cognitive reserve is one of the main factors driving this trend, through a rise in educational attainment. After the Second World War, there was an increase in schooling of about an extra year of education across the US population and in other countries. Research suggests that people with more education, or those who learn a new language or learn to play a musical instrument, may be resilient to symptoms of dementia.

That doesn't mean they escape the ravages of vascular dementia or the plaques of Alzheimer's, but they may cope better with the damage. The idea is that by challenging your brain during education, you create a more fit brain that can compensate for problems that you may have as you age.

Increased cognitive reserve is thought to help in two ways: boosting the brain's ability to work around damaged areas and promoting more efficient processing. That might also explain why people with more education seem to decline so rapidly: it's not that Alzheimer's comes on suddenly, it's that by the time symptoms manifest the disease may already be quite advanced.

The other main factor behind the fall in dementia rates is better control of cardiovascular issues. While the prevalence of conditions such as high blood pressure and diabetes has risen over the years, there has also been an increase in treatments that can limit their damage.

Making a difference

It is important to acknowledge that much of dementia risk is down to genetics, about 70 per cent in the case of Alzheimer's disease. Often patients lament that they didn't do enough, but sometimes there is only so much you can do. Still, if 30 per cent or more of dementia risk is down to lifestyle and environmental factors, there is an opportunity to make a difference. Maintaining social connections and a healthy diet, exercising regularly, practising good sleep habits and pursuing intellectual challenges may all delay or lessen symptoms of dementia later in life. Building a cognitive reserve is a lifetime's enterprise. The earlier you start, the better, as these changes have the most effect when they are started at a younger age.

It's never too late to start either. Fighting senility with mental gymnastics has become part of the anti-ageing folklore, and there is now good evidence that mental activity really does cushion people against age-related decline.

Mental activity is not the only thing that helps. Physical activity is crucial too. Regular exercise not only addresses risk factors such as obesity and cardiovascular health, but it increases the creation of brain cells, connections between neurons, and production of nerve growth factors and neurotransmitters. You don't have to run ultra-marathons to reap the benefits. Just an hour-long walk a few times a week can make a difference.

EAT YOUR WAY TO DEMENTIA

Junk food addicts take note: a high calorie diet isn't just bad for your body, it may also trigger Alzheimer's disease. Type 2 diabetes is a risk factor for Alzheimer's, but there is growing evidence that the link between the two diseases could be stronger. Rats fed so that they develop diabetes have brains littered with amyloid plaques – one of the calling cards of Alzheimer's. Findings such as these have led some researchers to wonder whether the disease may sometimes be another version of diabetes – one that hits the brain. Some have even dubbed it 'type 3 diabetes'. If they are right, the implications are troubling. Since calorific foods are known to impair our body's response to insulin, we may be unwittingly poisoning our brains every time we chow down on burgers and fries.

EXERCISE TO HELP
YOUR MIND

PUMPING IRON TO sculpt your biceps. Yoga poses to stretch and relax. Running to whittle your waistline and get fit fast. There are plenty of reasons why it's smart to exercise. Here's another: physical exercise makes your brain work better.

Exercise, particularly aerobic exercise, gets the blood pumping, bringing more oxygen, hormones and nutrients to your brain. This makes the brain – like your muscles, lungs and heart – grow stronger and more efficient. There is more to it than that, though. Studies, first in mice, and more recently in humans, have shown that aerobic exercise stimulates the neurons in the hippocampus – a region important for memory – to pump out a protein called brain-derived neurotrophic factor (BDNF), which promotes the growth of new neurons. Adult exercisers who had the highest levels of BDNF in their blood had bigger increases in the size of their hippocampus than control groups.

Intriguingly, different forms of exercise seem to affect the brain in different ways. Strength-training stimulates the release of another molecule: insulin-like growth factor-1 (IGF-1) from the liver. IGF-1 increases communication between brain cells and promotes the growth of new neurons and blood vessels.

Lifting weights also decreases levels of homocysteine, an inflammatory molecule that has been found to rise in the brains of older adults with dementia.

Regular exercise is particularly important for ageing brains. People who exercise at least twice a week in middle age are less likely to develop dementia in old age, for example. The earlier we start, the better. A study of children aged five to fourteen in New York City found that students in the top 5 per cent of the fitness rankings scored 36 percentile points higher on standardised academic tests than students ranked in the bottom 5 per cent. Similarly, an analysis of the medical records of over 1 million Swedish men who enrolled for military service as teenagers found that changes in fitness between the ages of fifteen and eighteen seemed to correlate with their intelligence scores and cognitive abilities at the end of that time period.

The right workout

So how to choose the right exercise for the mental benefits you want? For kids, it's a no-brainer: just let them run around. Studies show that even a twenty-minute walk has immediate effects on children's attention, executive function and achievement in mathematics and reading tests. It's unclear how long the effect lasts, however, so taking 'movement' breaks every couple of hours might be a good way to keep them on top form.

Other studies suggest that coordination exercises, such as basketball, volleyball or gymnastics, over the course of five months, helped children do better in tests that required concentration and ignoring distractions. This could be linked to changes in the cerebellum – the finely wrinkled structure at the base of the brain involved in coordinating movement but also known to have a role in sustaining our attention.

In general, the best exercises seem to be those that combine moving and thinking, challenging not only our sense of proprioception (tracking the position and orientation of the body) but also other elements, such as navigation or calculation. This could be as simple as doing the things we enjoyed as children. In one study, adults who were asked to spend two hours climbing trees, crawling along beams and running barefoot, subsequently showed improvements in their working memory, the ability to hold information in the memory and also to manipulate 'this information'. Other possibilities are sports such as surfing and climbing that require not only balance and strength but also planning and accurate movement.

Habits and behaviours

As well as these cognitive benefits, exercise may also help change our habits and behaviours. High-intensity interval training (HIIT), which involves quick spurts of all-out exercise, has been found to reduce food cravings immediately after working out and for a day or so afterwards. This might be linked to the 'hunger hormone', ghrelin, which tells the brain that the stomach is empty and is at its lowest levels after exercise. Research suggests that even less-intense bouts of activity can reduce cravings for both sugary snacks and cigarettes. Just a fifteen-minute walk or cycle ride is enough to reduce activity in brain regions implicated in addiction, making willpower less of a factor.

Then there is the fact that any exercise, even a leisurely stroll, can release stress. Since stress can inhibit your brain's responses when solving a problem, it stands to reason that removing this block might enable more ideas to form. After practising yoga for eight weeks, highly stressed adults not only felt less stressed but brain scans showed shrinkage in the amygdala, the structure in the brain that processes anxiety and other emotions.

Overall, though, it doesn't matter what kind of exercise you do as long as you do something. Comparisons of the brains of elderly people who exercised, or not, throughout their lives, showed not only that the exercisers' brains appeared ten years younger than the couch potatoes, but their cognitive skills declined more slowly. Perhaps the best form of exercise is the one that you can stick at for the life of your brain.

MIND GAMES FOR MOBILITY

Moving the body can tone the mind, and there is evidence that the reverse is also true. Walking and keeping a steady posture are automatic, but still require a certain amount of attention. Both attention and physical agility tend to fade with age, and seem to be linked. Older people who score poorly on attention tests are more likely to take a fall. In one study, elderly people played a computer game designed to tax their attention skills for five weeks and monitored their walking speed, an indicator of the risk of falling (the slower the walk, the higher the risk – it's to do with the ability to integrate multiple sources of information such as keeping balance, avoiding bumps and thinking about where you're going). Walking speed increased after training, suggesting that working their minds had limbered up their bodies too.

HOW TO UNLEASH
YOUR INNER GENIUS

WHAT DAY OF the week was 5 June 1963? Chances are you can't say without looking it up, but a calendar savant could name it in an instant.

It's just one of many uncanny abilities demonstrated by savants, who are often, but not always, autistic. Kim Peek, who inspired the 1988 film *Rain Man*, memorised over 8,600 books and can name all the US telephone area codes. Artist Stephen Wiltshire can draw an entire landscape after seeing it just once. These talents are all the more striking because they often occur in people whose intelligence is otherwise limited.

Origins of savant skills

Psychologists have long been fascinated by savant skills. The orthodox view is that these 'islands of genius' result from obsessive use of a particular mental capacity. Some neuroscientists have a different view: that we all have the potential for these exceptional skills, which lie dormant within our subconscious.

The main reason for thinking everyone possesses untapped savant skills is that they can appear spontaneously after brain damage. One case in the literature documents how a child suddenly acquired spectacular calendar-calculating skills and an extraordinary memory for dates and music following a head injury at the age of ten. The striking thing is that this patient, and others with 'acquired savantism', had damage in the same area of the brain, the left frontotemporal lobe. Autism is associated with deficits of the left hemisphere too.

Why would deficits in this part of the brain lead to unusual talents? It's too simplistic to say that the right brain and left brain are neatly divided, but the hemispheres do specialise in certain functions. The left side deals with language and logical thinking, while the right hemisphere tends to govern creativity and our more artistic side. When the left side is suppressed, this can allow the right side to step in to compensate.

A key proponent of this view is neuroscientist Allan Snyder. His version of this model is that your unconscious brain extracts all the raw sensory details about the world around you – the tones, pitches, lines, light and shadow. This information is far more than we can deal with, but it is where, he believes, we experience the world as it really is.

Except most people never see this version of events. Our unconscious mind takes the flood of information and simplifies and categorises it into manageable packages. Where the unconscious sees lines and patterns of dark and shade, our conscious mind might see a horse. This is an efficient way for our minds to work because it allows us to spot things quickly, to name them and communicate the ideas. Snyder believes that savants experience only the raw sensory information of the unconscious. The reason most people don't experience the world this way is because of interference from higher order cognitive processing. By switching this off, extraordinary skills could be switched on.

It was a controversial idea, so Snyder set about trying to prove it using 'transcranial magnetic stimulation' (TMS) in which a strong magnetic field is placed on the scalp to temporarily halt activity in the nearby part of the brain (this might sound drastic, but TMS is routinely used in neurology departments and hospitals.) He focused TMS on the left frontotemporal lobe to see if savant skills emerged when this brain region was inhibited. First, he looked at the effect of TMS on drawing skills. It didn't work on everyone, but some volunteers showed a major change in their drawing style. Another study showed that after TMS many participants were able to solve a notoriously challenging puzzle, which had previously defeated them. Other studies show that memory, mathematics and calendar counting skills can be switched on by switching off the left frontotemporal lobe.

Electrical stimulation

Another way to externally manipulate brain activity is through transcranial direct current stimulation (tDCS), in which electrodes are applied to the scalp to influence neuron activity. Targeting tDCS at the brain's right parietal lobe can boost arithmetic ability. The technique can also improve insight. When it was applied to volunteers' anterior frontal lobes – regions that play a role in how we perceive the world – they were three times more likely to complete a problem-solving task. Twenty minutes of tDCS to the brain's left perisylvian area speeds up and improves language learning.

Over the past decade thousands of studies have reported a beneficial effect of tDCS on behaviour and cognition. But recently, questions have been raised about the reliability and repeatability of these studies, so it's possible that the benefits of tDCS and TMS have been overhyped. The prospect of electrical 'thinking caps' for everyday use to give our neurons a boost is still up in the air.

Are there any other ways to unleash our brain's buried potential? Savant expert, psychiatrist Darold Treffert – who, like Snyder, thinks that there is dormant potential within all of us – recommends a low-tech route. To release creative potential he suggests spending time 'rummaging in your right hemisphere' – taking up new hobbies, broadening your interests, anything that involves creativity and new learning other than using our well-worn, logical, language-dependent left hemisphere.

OBSESSIVE PRACTICE

Savants practise tirelessly, but how much does this contribute to their specific talent? Could we all learn these amazing skills if only we practised more? A study of calendrical savants examined this issue. It showed that their extraordinary talent didn't seem to depend on any abnormal cognitive processes or a fundamentally different brain. Indeed, when one of these savants was asked to learn a new calendrical system with a limited amount of time to practise, he performed no better than a non-savant. The conclusion was that these savant skills developed from intense practice and are something we could all develop if we put in the hours. But how many of us have the motivation? In fact, the remaining mystery is not so much how savants achieve their talents, but what drives them in the first place. Motivation is an enormous driving force in giftedness and in savants, but we don't know a lot about it.

BRAIN GYMS: IS IT WORTH VISITING?

PICTURE THIS: YOU'RE concentrating hard, staring at a small white square in the middle of a computer screen. Any second now a letter is going to flash up inside the box. At the same time a bird will pop up elsewhere on the screen. Your task is to hit the bird with your mouse, then type the letter in the box. But this isn't any ordinary computer game. It's brain training. The more you practise, the better you'll get and the more powerful your brain will become. That's the theory at least. Does it actually work?

Cognitive games

Brain training is based on the well-established scientific fact that our brains are 'plastic', they change and adapt in response to learning challenges. In the early 2000s, commercial brain games started to emerge, promising to tax key cognitive skills and to keep our brains sharp as we age. Most feature the kinds of cognitive games and tests that psychologists use to test key mental skills. Unlike computer games designed purely for entertainment, brain-training games are meant to be adaptive, adjusting challenge levels in response to a player's changing performance. The thinking is that this should gradually improve a player's memory, attention, focus and multitasking skills.

It sounded good, but while the brain-training industry grew into a multimillion-dollar business, scientific studies of the effects of the games began to cast doubt on their effectiveness. While some trials showed success, they were criticised for being too small to produce meaningful results. A trial of online brain-training games involving over 11,000 volunteers in 2010 concluded that brain training made no difference to their general cognition.

The inability to transfer benefits to daily life is the main problem with generic brain games. It is easy enough to show that a person playing a particular game gets better at the game itself, but proving that a trained skill transfers to better performance at a different task is fiendishly difficult. It doesn't matter how good you become at a particular memory game, for instance, it doesn't necessarily mean that you will find it easier to remember your shopping list. So far, no large published trial has yet shown concrete evidence that brain training has an effect on life skills.

Worse, studies that have pitted brain games against regular computer games found no evidence that specific brain games are any better at improving your brain skills than any other kind of game. While people who played brain games did show improvements in some cognitive skills such as attention and focus, so did those who played other computer games, and the people who played no games at all. This suggests they all performed better on the second batch of tests simply because they'd already done them once. This practice effect, combined with the fact that our brains are changing all the time to adapt to the challenges of everyday life, makes it very difficult to prove that any improvement in score has anything to do with playing a game.

In 2016, the backlash peaked when one brain-game provider, Lumos Labs, was ordered to pay $2 million by the US Federal Trade Commission for false advertising. The ruling found that adverts claiming that the company's memory and attention games could reduce the effects of age-related dementia and stave off Alzheimer's disease were misleading and not backed up by scientific evidence.

Hidden benefits?

Given all of this, it doesn't seem worth the time, effort and money to play brain games in the hope that they will sharpen your brain or prevent age-related decline. On the other hand, it's possible that there are small improvements for some people – which are diluted in large studies where scores get averaged – or that the games currently available are too blunt a tool to get the kinds of improvements they are aiming for. Some forms of specific cognitive training show promise in treating the cognitive deficits associated with schizophrenia and in slowing the progression of Alzheimer's.

A study of 2,800 people over the age of sixty-five, for instance, found that those who did a type of brain training intended to boost brain processing speed were 29 per cent less likely to develop dementia over a ten-year period. The processing training involved identifying objects briefly displayed on a computer screen. As trials go on, the objects are shown for shorter amounts of time, among other distracting objects, and with increasingly detailed backgrounds. Interestingly, it was only brain processing speed training that had this beneficial effect. Participants who only completed training for memory or reasoning were just as likely to develop dementia as the control group.

There are limitations to the study, including the fact that dementia was determined by self-reporting or cognitive assessments, not a full clinical diagnosis, and some scientists remain sceptical that a simple intervention could have such dramatic effects. More evidence is needed before many are convinced that brain training really can prevent dementia. For now, the jury is still out on whether brain training is a waste of time, or just needs to be adapted into a form that does what it says on the tin.

MUSICAL MUSCLE

Our brains constantly adapt to outside influences, but some things make a bigger impression than others. Music is one of them. The main brain changes in musicians concern the regions involved in processing sounds and fine movements. These areas are also involved in understanding speech and language, so it is perhaps not surprising that musicians are better at learning to use and recognise the sounds of foreign languages. This skill may also enhance empathy, because it fine-tunes a person's ability to recognise emotional nuances in speech.

There is also evidence that professional pianists are much better than non-musicians at discriminating two closely separated points, perhaps from years of sight-reading. They also improved faster with practice, suggesting that music makes brains more plastic in general. Learn an instrument, then, and it might get easier to learn everything else.

HACK YOUR BRAIN
THE HIGH-TECH WAY

THERE ARE MANY ways to pimp the brain using technology – some more out-there and risky than others.

Neurofeedback

This works by visualising real-time brain activity (measured by EEG) and then changing this activity through thought. Neurofeedback shows promise as a treatment for depression, obsessive–compulsive disorder and schizophrenia. It may also teach users how to regulate their brain function, enhancing cognitive abilities such as spatial memory and planning.

Blue light

The blue light emitted by mobile phone screens gets a bad press for disrupting sleep, but there is a flipside: this alertness can be helpful for undertaking cognitive tasks. Tests show that logical reasoning, maths and reaction time all improved after blue light exposure.

Smart drugs

Many of us already get a mild cognitive boost from caffeine and nicotine, but some use prescription drugs in ways that doctors would not approve. Modafinil (the wakefulness drug prescribed for sleep disorders) is (mis)used to stay awake for long periods and Ritalin (prescribed for attention deficit disorder) aids concentration.

Transcranial direct current stimulation

In tDCS, electrodes on the scalp deliver a small current to nearby brain areas, changing neuronal activity (not to be confused with electroconvulsive therapy). Simple, cheap and safe, tDCS can boost mathematical skills, working memory, focused attention, creativity.

Transcranial magnetic stimulation

With TMS you don't need to break the skin to tinker with brain activity. Pulses from a strong electromagnet on the skull alter the activity of nearby neurons. Approved for the treatment of depression, TMS can also boost speed and accuracy in a range of cognitive tasks.

Deep brain stimulation

Electrodes implanted deep in the brain can have miraculous effects on the symptoms of Parkinson's disease and some mental illnesses. DBS can also affect mood, personality and even creativity – but the high risks of this invasive technology make cognitive enhancement a no-no.

MEDITATION: WHAT CAN IT DO FOR YOUR BRAIN?

THE PRACTICE OF meditation has thrown off the shackles of 1970s hippiedom and become something we perhaps all feel like we should be doing to take care of our brains. But what is the evidence that it is worth the time and effort?

In recent years, many studies have been asking the same question, and a picture has begun to emerge that, done regularly, meditation does have measurable effects on the brain. What those changes are seems to depend on the kind of meditation you do. There are many kinds of meditation, but in studies the most commonly used are either mindfulness meditation, which involves focusing attention on one aspect of consciousness (often the breath), or compassion-based meditations, which focus on projecting feelings of love and goodwill to other people.

Being mindful

Perhaps unsurprisingly, one of the most commonly found effects of focus-based meditation is an improved ability to focus. In brain-scanning studies, mindfulness meditation has been shown to increase thickness in the prefrontal cortex and parietal lobes, both of which are linked to attention control. Other experiments have found that after three months of meditation training volunteers showed a decreased 'attentional blink'. This is a measure of our moment-to-moment ability to re-focus and is measured by the time, in milliseconds, that it takes the brain to reset after one stimulus so that it can register another.

There are also signs that mindfulness meditation improves working memory, the capacity to hold in mind information needed for short-term reasoning and comprehension. Meditation often involves observing how sensory experience changes from moment to moment, and this could mean that, with practice, we could become better at holding fading sensory information in working memory so that we can keep up with the changing world a little more easily.

Since both focus and working memory are key cognitive skills that underlie most of what we do, these findings alone make it worth investing the time and effort to sit and breathe. Yet there may be other good reasons, too. Compassion-based meditation, which focuses on empathy, brings about brain changes not in key cognitive skills but in social and emotional ones. In one study, just seven hours of compassion meditation boosted altruistic behaviour and also made people feel happier. Brain circuits linked to empathy and the sharing of emotions – including the insula and the anterior cingulate cortex – have also been shown to be much more active in people who meditate regularly.

It could also help us grapple with our own emotions. Brain scans taken after a three-month meditation course showed areas involved in the limbic system, which processes emotions, and the anterior insula, which helps bring emotions into conscious awareness, were increased in size. The amygdala, a part of the limbic system that processes fear and emotional memories, seems to be particularly affected, with lower levels of activity during meditation and shrinkage over time. This in particular could account for meditation's oft-touted ability to reduce stress and anxiety.

What's more, there is some evidence that this ability to lower stress might have knock-on effects that improve our physical health. Recent research suggests that 'mind–body practices' including meditation dampen the activity of genes associated with inflammation, part of the immune system that can damage the body if it is switched on for long periods. Chronic inflammation is thought to be an important route by which psychological stress increases a person's risk of developing disease and is associated with increased risk for psychiatric disorders, autoimmune conditions such as asthma and arthritis, cardiovascular disease, neurodegenerative disease and some types of cancer. A finding that a key protein, which acts as an inflammation 'on-switch', is down-regulated in people who meditate suggests that meditation might be one way to reduce the effects of stress.

The dark side

Yet while all this evidence seems positive, not all scientists agree that we should rush to the nearest retreat. A small number of studies have reported serious side effects from meditation. One reported that 7 per cent of people who attended a meditation retreat suffered panic and disorientation, hallucinations, terror, depression and even psychotic breakdown.

Others point out that happiness and well-being are not what meditation was designed for. In both Buddhist and Hindu traditions, its purpose is to separate a person from their sense of self, which may bring up troubling feelings as well as making a person more susceptible to manipulation. Historians point out that the training of Japanese soldiers included the use of meditation techniques to ensure that the soldier lost his sense of self and 'became' the very order he received.

Some psychologists are also concerned that encouraging people to become detached from negative thoughts might not always be beneficial. Some negative thought patterns need to be challenged and changed, not accepted and lived with. In short, while there is mounting evidence that meditation can change the mind in powerful and life-changing ways, it may have a dark side that is worth bearing in mind. However tempting it may be to train your brain to focus better and be less stressed, like all brain-altering experiences, even meditation comes with a health warning.

FIND YOUR FLOW

Have you ever felt a feeling of Zen-like, effortless concentration while carrying out an activity, as though time seems to stop as you focus your mind completely on a task? If you have, then you've experienced an elusive mental state known as 'flow'. This experience crops up repeatedly when experts describe what it feels like to be at the top of their game, but beginners can experience it too – indeed, it's linked to better progress. Some of us are more naturally predisposed to the flow state than others. The feeling is linked to a reduction of activity in the prefrontal cortex, typically associated with higher cognitive processes. This fits in with the idea that better learning comes when you turn off conscious thought.

HOW TO MASTER
YOUR MEMORY

MEMORY IS A wonderful gift, but it has its pitfalls. Who hasn't felt facts slip through their mind like sand through a sieve as they crammed for an exam? At other times, forgetting may be the difficulty, as we struggle to banish memories of painful events. Thankfully, a growing understanding of the human mind offers many ways to help you make the most of your innate abilities.

Hit the sweet spot

When trying to memorise new material, you may find yourself staring endlessly at the page in the hope that its contents will somehow seep into your mental vault. One of the most effective ways of learning for an exam, though, is to test yourself repeatedly, which may be simpler to apply to your studies than other, more intricate methods, such as the formal mnemonic techniques used by expert memorisers.

It's important to pace yourself, too, by revisiting material rather than cramming it all in during a single session. When doing so, you should make the most of sweet spots in the timing of your revision. If you are studying for an exam in a week's time, for instance, you will remember more if you leave a day or so between your first and second passes through the material.

Limber up

Besides keeping your body – and therefore your grey matter – in generally good shape, a bit of exercise can offer immediate benefits for anyone trying to learn new material. In one study, students taking a 10-minute walk found it much easier to learn a list of thirty nouns, compared with those who sat around, perhaps because it helped increase mental alertness.

Short, intense bursts of exercise may be the most effective. In one experiment, participants learning new vocabulary performed better if their studies came after two 3-minute runs, as opposed to a 40-minute gentle jog. The exercise seemed to encourage the release of neurotransmitters involved in forming new connections between brain cells.

Make a gesture

There are also more leisurely ways to engage your body during learning, as the brain seems to find it easier to learn abstract concepts if they can be related to simple physical sensations. Various experiments show that acting out an idea with relevant hand gestures can improve later recall, whether you are studying a foreign language or memorising the rules of physics.

It may sound dubious, but even simple eye movements might help. People are better able to remember a list of words they have just studied if they repeatedly look from left to right and back for 30 seconds straight after reading the list – perhaps because it boosts the transfer of information between the two brain hemispheres. It's worth noting, however, that this only seems to benefit right-handers. Perhaps the brains of left-handed and ambidextrous people already engage in a higher level of cross-talk, and the eye-wiggling only distracts them.

Engage your nose

Often it's not just facts that we would like to remember, but whole events from our past as we reminisce about the good ol' days. Such nostalgia is not just an indulgence – it has been linked to a raft of benefits, such as helping us to combat loneliness and feelings of angst. If you have trouble immersing yourself in your past, you could borrow a trick from Andy Warhol. He used to keep a well-organised library of perfumes, each associated with a specific period of his life. Sniffing each bottle reportedly brought back a flood of memories from that time – giving him useful reminders whenever he wanted to reminisce. Warhol's approach finds support in a spate of recent studies showing that odours tend to trigger particularly emotional memories, such as the excitement of a birthday party; they are also very effective at bringing back memories from our childhood. Some have even suggested that you could boost your performance in a test by sniffing the same scent during your revision and on the day of the exam.

Oil the cogs

Everyone's memory fades with age, but your diet could help you to keep your faculties for longer. You would do well to avoid high-sugar fast foods, for instance, which seem to encourage the build-up of the protein plaques characteristic of Alzheimer's disease.

In contrast, diets full of flavonoids, found in blueberries and strawberries, and omega-3 fatty acids, found in oily fish and olive oil, seem to stave off cognitive decline by a good few years – perhaps because the antioxidants protect brain cells from an early death.

LEARNING TO FORGET

Sometimes we are haunted by unwanted memories: a moment of embarrassment perhaps, or a painful break-up. Banishing such recollections from our thoughts is difficult, but there may be ways of stopping fresh memories of painful events from being consolidated into long-term storage in the first place. For example, in one study subjects were asked to watch a disturbing video, before engaging in various activities. Those playing the video game *Tetris* subsequently experienced fewer flashbacks to unpleasant scenes in the film than those taking a general knowledge quiz, perhaps because the game occupied the mental resources usually involved in cementing memories. Playing relaxing music to yourself after an event you would rather forget also seems to help, possibly because it takes the sting out of the negative feelings that normally cause these events to stick in our minds.

ANSWERS

Test your lateral thinking, page 146

1. USHER (Us, She, He, Her)
2. The sentence contains no E
3. S, S (standing for SIX and SEVEN)
4. HEadacHE and HEartacHE
5. 6/9
6. Make a triangular-based pyramid
7. WATER (as in H_2O or H to O)
8. They are ordered alphabetically, according to the letter the number begins with
9. They are grandfather, father and son
10. DNA (this molecule, which carries genetic instructions, is constructed from four different molecules called nucleotides referred to by the letters A, T, G and C)
11. N (all these capital letters are the same when rotated 180 degrees)
12. Yes
13. Bras, millionaires, princes
14. She is a midwife
15. Short

16. All the words turn into another word if you drop the first letter
17. Each word is one letter longer than the preceding word

Sources: Morton Schatzman, Futility Closet

How good is your memory?, page 196

Here are the average scores for *New Scientist* employees. Did you beat them?

1. Language. 7/10
2. Names. 12/15
3. Signs. 7/8
4. Numbers. 14/20
5. Cards. 8/9

Are you a psychopath?, page 221

If you scored highly on the psychopath test, what does this mean?

For psychologists, a psychopath is someone with a distinct cluster of personality traits including charm, charisma, fearlessness, ruthlessness, narcissism, persuasiveness, and lack of conscience. Sure, these traits may well come in handy if you aspire to be an axe-murderer. But they can also come in handy in the courtroom, on the trading floor, or in an operating theatre. It just depends on what else you've got going on in your personality, and the start you get in life, says psychologist Kevin Dutton.

Another misconception about psychopaths is that it's very black and white: you're either one or you're not. In fact, psychopathy – like height, weight and IQ – lies on a spectrum. Yes, at the sharp end you may well find your serial killers and axe murderers. But all of us have our place at some point along the continuum. Some of us may score higher on some psychopathic traits than on others. But unless you score high on all of them, you don't really have anything to worry about

Test your creative spark, page 261

1. See below
2. There is no right answer
3. An egg
4. Note
5. Move the circles at the corners. See below

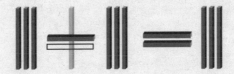

GLOSSARY

Amygdala: The fight-or-flight centre of the human brain, responsible for gut reactions, motivation, emotion and fear

Autobiographical memory: Our knowledge of experiences we had at a specific time and place, such as a particular train journey, combined with general knowledge of ourselves

Axon: A long, slender nerve fibre that shuttles electrical impulses away from the main body of a neuron

Brainstem: A primitive part of the brain that is responsible for many basic functions, such as heart rate and breathing. It also regulates the central nervous system and the sleep cycle

Cerebellum: This brain region's main job is to coordinate voluntary movements and balance, as well as proprioception – our awareness of our own bodies. It is also thought to be involved in speech and our ability to learn specific motor actions

Cerebrum: This is the largest part of our brain, made up of two cerebral hemispheres. It is a collection of brain structures, including the cerebral cortex and the hippocampus. The cerebrum controls our personality, hearing and vision. Together with the cerebellum, it also controls voluntary actions

Cortex: Made up of grey matter, the cortex is the thin, highly folded outer layer of the brain. It coordinates some of our most advanced mental functions, like planning, language and complex thoughts

Default mode network: This is the brain's autopilot mode. It is characterised by a pattern of brain activity that is seen when a person is at rest or allows their mind to wander, but diminishes when they begin to concentrate on a task

EEG: Stands for electroencephalography, a method of recording the brain's electrical signals via electrodes placed on the scalp

Episodic memory: Recollections of particular events we have experienced (a component of autobiographical memory)

Grey matter: This is mainly comprised of the bodies of cells called neurons. The brain's cortex is made from this kind of tissue

fMRI: Stands for functional magnetic resonance imaging. This kind of scan can measure blood flow in the brain to show which parts of the brain are active when a person carries out a particular task

Frontal lobe: This brain region has many functions relating to cognition and behaviour. It plays a large role in decision-making, impulses and conscience, as well as integrating memories from across the brain

Heuristic: A mental shortcut or 'rule of thumb' that allows us to make fast decisions with minimal cognitive effort

Hippocampus: The brain region where memories are stored and processed. Humans and other mammals have two hippocampi, one in each side of the brain

MRI: Stands for magnetic resonance imaging. MRI scanners use a strong magnetic field and radio waves to build a 3D picture of a body part and its internal structure

Myelin: The fatty substance that covers the fibre-like axons of nerve cells. A myelin sheath helps signals to travel faster through an axon

Nerve impulses: These are electrical signals that travel along nerve fibres, allowing neurons to communicate with each other

Neurons: These tree-like nerve cells are the fundamental building blocks of the brain. They transmit information using electrical and chemical signals

Neurotransmitter: This is a molecule that carries signals across the gaps between neurons. The brain has many different neurotransmitters such as GABA, serotonin, dopamine and glutamate

Prefrontal cortex: Located at the front of the brain, the prefrontal cortex is involved in planning and other complex behaviours, like personality and considering things from another person's perspective

Procedural memory: A type of unconscious memory for knowing how to do things, such as tying shoelaces or riding a bike

REM sleep: Rapid eye movement sleep is a phase of sleep in which our muscles relax and we have vivid dreams. REM sleep is thought to be important for consolidating memories

Semantic memory: Knowledge of facts, such as the meanings of words or Paris being the capital of France

Short-term memory: These are memories we hold at the front of our mind for a short period of time, such as a telephone number we are about to use

Striatum: A brain region important for movement, mood and reward

Synapse: The gaps or chemical junctions between neurons.

Electrical and chemical signals between neurons pass across these gaps

Working memory: This enables the manipulation of material in your short-term memory. Your short-term memory might help you to remember what someone has just said to you, but your working memory would allow you to recite it to them backwards or pick out the first letter of each word

White matter: This is the branching network of thread-like tendrils that spread out from the bodies of nerve cells to connect them to other nerve cells

FURTHER READING

Meet Your Brain

30-Second Brain: The 50 Most Mind-Blowing Ideas in Neuroscience, Each Explained in Half a Minute, Anil Seth (Icon Books, 2014)

Inventing Ourselves: The Secret Life of the Teenage Brain, Sarah-Jayne Blakemore (Doubleday, 2018)

Behave: The Biology of Humans at Our Best and Worst, Robert Sapolsky (Penguin Random House, 2017)

Do No Harm: Stories of Life, Death and Brain Surgery, Henry Marsh (W & N, 2014)

The Language Myth: Why Language is Not an Instinct, Vyvyan Evans (Cambridge University Press, 2014)

The Future of the Mind: The Scientific Quest to Understand, Enhance and Empower the Mind, Michio Kaku (Random House, 2014)

The Prehistory of the Mind: A Search for the Origins of Art, Religion and Science, Steven Mithen (Thames & Hudson, 1996)

Perception

Your Brain is a Time Machine: The Neuroscience and Physics of Time, Dean Buonomano (W. W. Norton, 2017)

Surfing Uncertainty: Prediction, Action, and the Embodied Mind, Andy Clark (Oxford University Press, 2016)

The Mind's Eye, Oliver Sacks (Knopf, 2010)

See What I'm Saying: The Extraordinary Powers of our Five Senses, Lawrence D. Rosenblum (W. W. Norton, 2010)

The Universal Sense: How Hearing Shapes the Mind, Seth Horowitz (Bloomsbury, 2012)

Hallucinations, Oliver Sacks (Knopf, 2012)

Intelligence

Are We Getting Smarter?: Rising IQ in the Twenty-First Century, James Flynn (Cambridge University Press, 2012)

Does Your Family Make You Smarter? Nature, Nurture, and Human Autonomy, James Flynn (Cambridge University Press, 2016)

Outliers: The Story of Success, Malcolm Gladwell (Little, Brown and Company, 2008)

Consciousness

Self Comes to Mind: Constructing the Conscious Brain, Antonio Damasio (Pantheon, 2010)

Phi: A Voyage from the Brain to the Soul, Giulio Tononi (Pantheon, 2012)

Soul Dust: The Magic of Consciousness, Nicholas Humphrey (Quercus, 2011)

Why Red Doesn't Sound Like a Bell: Understanding the Feel of Consciousness, J. Kevin O'Regan (Oxford University Press, 2011)

Consciousness and the Brain: Deciphering How the Brain Codes our Thoughts, Stanislas Dehaene (Viking Press, 2014)

The Unconscious

Drunk Tank Pink: The Subconscious Forces That Shape How We Think, Feel And Behave, Adam Alter (Penguin, 2013)

Incognito: The Secret Lives of the Brain, David Eagleman (Pantheon Books, 2011)

The Wandering Mind: What the Brain Does When You're Not Looking, Michael Borballis (University of Chicago Press, 2015)

Thinking

Thought: A Very Short Introduction, Tim Bayne (Oxford University Press, 2013).

Intuition Pumps and Other Tools for Thinking, Daniel Dennett (W. W. Norton & Company, 2013)

Thinking, Fast and Slow, Daniel Kahneman (Farrar, Straus and Giroux, 2011)

Thinking and Reasoning: A Very Short Introduction, Jonathan Evans (Oxford University Press, 2017)

Mindware: Tools for Smart Thinking, Richard Nisbett (Allen Lane, 2015)

Gödel, Escher, Bach: An Eternal Golden Braid, Douglas Hofstadter (Basic Books, 1979)

Memory

The Memory Illusion: Remembering, Forgetting, and the Science of False Memory, Julia Shaw (Random House, 2017)

Pieces of Light: The New Science of Memory, Charles Fernyhough (Profile Books, 2012)

Moonwalking with Einstein: The Art and Science of Remembering Everything, Joshua Foer (Penguin Press, 2011)

The Self

The Man who Wasn't There: Investigations Into The Strange New Science Of The Self, Anil Ananthaswamy (Penguin, 2016)

Who's in Charge?: Free Will and the Science of the Brain, Michael Gazzaniga (Ecco Press, 2011)

Free Will, Sam Harris (Free Press, 2012)

The Mind Club: Who Thinks, What Feels, and Why It Matters, Daniel Wegner and Kurt Gray (Viking, 2016)

Creativity

The Eureka Factor: AHA! Moments, Creative Insight and the Brain, John Kounios and Mark Beeman (Random House, 2015)

Eureka!: Discovering your Inner Scientist, Chad Orzel (Basic Books, 2014)

How to Fly a Horse: The Secret History of Creation, Invention and Discovery, Kevin Ashton (Doubleday, 2015)

The Cambridge Handbook of the Neuroscience of Creativity, Rex Jung, Oshin Vartanian (Cambridge University Press, 2018)

Decision-Making

The Enigma of Reason: A New Theory of Human Understanding, Dan Sperber and Hugo Mercier (Allen Lane, 2017)

Risk Intelligence: How to Live with Uncertainty, Dylan Evans (Simon & Schuster, 2012)

Nudge: Improving Decisions About Health, Wealth, and Happiness, Richard Thaler and Cass Sunstein (Yale University Press, 2008)

My Brain Made Me Do It: The Rise of Neuroscience and the Threat to Moral Responsibility, Eliezer Sternberg (Prometheus, 2010)

The Social Brain

Mindwise: Why We Misunderstand What Others Think, Believe, Feel, and Want, Nicholas Epley (Knopf, 2014)

Wired for Culture: Origins of the Human Social Mind, Mark Pagel (W. W. Norton & Company, 2012)

The Domesticated Brain, Bruce Hood (Pelican, 2014)

Sleep and Dreaming

Why We Sleep: The New Science of Sleep and Dreams, Matthew Walker (Allen Lane, 2017)

Rest: Why You Get More Done When You Work Less, Alex Soojung-Kim Pang (Basic Books, 2016)

Sleep: A Very Short Introduction, Steven Lockley and Russell Foster (Oxford University Press, 2012)

Troubleshooting

Unthinkable: An Extraordinary Journey Through the World's Strangest Brains, Helen Thomson (John Murray, 2018)

Angst: Origins of Anxiety and Depression, Jeffrey Kahn (Oxford University Press, 2012)

Anatomy of Love: A Natural History of Mating, Marriage, and Why We Stray, Helen Fisher (W. W. Norton, 2016)

The Man Who Mistook His Wife for a Hat and Other Clinical Tales, Oliver Sacks (Summit, 1985)

Homo Deus: A Brief History of Tomorrow, Yuval Noah Harari (Harvill Secker, 2016)

Making the Most of It

Override: My Quest to Go Beyond Brain Training and Take Control of My Mind, Caroline Williams (Scribe, 2017)

The Science of Meditation: How to Change Your Brain, Mind and Body, Daniel Goleman, Richard Davidson (Penguin, 2017)

The Buddha Pill: Can Meditation Change You?, Miguel Farias and Catherine Wikholm (Watkins Publishing Limited, 2015)

Rainy Brain, Sunny Brain: How to Retrain Your Brain to Overcome Pessimism and Achieve a More Positive Outlook, Elaine Fox (Basic Books, 2012)

Superhuman: Life at the Extremes of Mental and Physical Ability, Rowan Hooper (Little, Brown, 2018)

What Makes Your Brain Happy and Why You Should Do the Opposite, David DiSalvo (Prometheus Books, 2012)

ACKNOWLEDGEMENTS

THIS BOOK WOULD not exist without the ideas, hard work and support from the *New Scientist* family. Huge thanks are due to Graham Lawton who devised the concept, came up with many ideas and wrote some of the articles; to Caroline Williams and Helen Thompson whose excellent editing and writing skills were vital; to publisher John MacFarlane for his support, and to all the writers who have contributed their ideas and talent to the magazine.

For the original illustrated edition, enormous thanks are also due to the team at John Murray, especially Georgina Laycock and Kate Craigie for their wise guidance, brilliant ideas and enthusiasm. Thanks also to Nick Davies, Will Speed for the cover and Mandi Jones in production, Nicky Barneby for her design (and lots of cups of tea), Yassine Belkacemi and Jess Kim for publicity and marketing, Joanna Kaliszewska and Grace McCrum for rights, and Megan Schaffer, Lucy Hale and Megan Smith in sales.

The illustrated version of this book would also be nothing without the creative talent and huge commitment of illustrator Valentina D'Efilippo, who did an incredible job of turning ideas into illuminating graphics, always with a playful eye.

I would also like to thank psychotherapist Morton Schaztman for taking a trip down memory lane to articles he wrote for *New Scientist* back in 1983, in which he invited readers to solve lateral thinking problems in their sleep. Many of the questions in the lateral thinking quiz were devised by him. Thanks also to psychologist Kevin Dutton for letting us use his psychopath test, to neuroscientist Martin Pyka, and to Dave Johnson for his graphics skills.

On the home front, for their incredible support and encouragement, I would to thank Sasha, Isla, Eva, Susie, Sally, David, Lisa, Helen and especially Jonathan.

Some of the material in this book is adapted from articles previously published in *New Scientist*.

INDEX